International Conference "Energy Efficiency and Energy Saving in Technical Systems" (EEESTS 2020)

Held online due to COVID-19

IOP Conference Series: Materials Science and Engineering
Volume 900

Rostov-on-Don, Russia
16 - 17 June 2020

ISBN: 978-1-7138-2813-6
ISSN: 1757-8981

Printed from e-media with permission by:

Curran Associates, Inc.
57 Morehouse Lane
Red Hook, NY 12571

Some format issues inherent in the e-media version may also appear in this print version.

Institute of Physics
Dirac House, Temple Back
Bristol BS1 6BE UK

Phone: 44 1 17 929 7481
Fax: 44 1 17 920 0979

techtracking@iop.org

Additional copies of this publication are available from:

Curran Associates, Inc.
57 Morehouse Lane
Red Hook, NY 12571 USA
Phone: 845-758-0400
Fax: 845-758-2633
Email: curran@proceedings.com
Web: www.proceedings.com

TABLE OF CONTENTS

SECTION 1. MODERN ENERGY EFFICIENT AUTOMATION TECHNOLOGY

IOP Conf. Series: Materials Science and Engineering **900** (2020) 011001 doi:10.1088/1757-899X/900/1/011001

II International conference
"Energy efficiency and energy saving in technical systems" (E^3STS-2019)

Conference was held at June 16-17, 2020, in Rostov-on-Don, Russia, on the base of Don State Technical University.

Due to the difficult epidemiological situation associated with COVID-19, the conference will be held online using ZOOM technology.

The main direction of the conference is to improve international mobility and cooperation between the Don State Technical University and foreign universities participating in the conference: Le Quy Don University of Science and Technology (Hanoi, Vietnam); University of Kragujevac (Čačak, Serbia); Tehran University (Tehran, Iran), Rutgers State University, (New Brunswick, New Jersey, USA), Sogang University (Seoul, South Korea), Ministère Équipements Transport Logistique Eau (MEQTLE, Rabat, Maroc) and others.

The conference is held in the form of two sections devoted to highlighting the achievements of energy efficiency and energy saving indicators through the introduction of automation systems (section 1) and through the use of modern and efficient data processing and measurement algorithms (section 2).

In 2020, over 160 delegates from 7 countries of the world & 22 institutions took part in the online conference.

Peer review declaration

All papers published in this volume of IOP Conference Series: Materials Science and Engineering have been peer reviewed through processes administered by the Editors. Reviews were conducted by expert referees to the professional and scientific standards expected of a proceedings journal published by IOP Publishing.

- **Type of peer review: Single-blind / Double-blind / Triple-blind / Open / Other (please describe):** Open review
- **Conference submission management system:** google forms
- **Number of submissions received:** 24 papers
- **Number of submissions sent for review:** 24 papers
- **Number of submissions accepted:** 21 papers
- **Acceptance Rate (Number of Submissions Accepted / Number of Submissions Received X 100):** 87.5%
- **Average number of reviews per paper:** two reviews
- **Total number of reviewers involved:** four persons
- **Any additional info on review process:** all articles were assessed for compliance with the IMRAD principle. The reviewer was also asked to answer a number of questions related to the quality of the content and design of the work. Thus, the review process was largely formalized and impersonal.
- **Contact person for queries:** Lukyanov Alexander, conf.dstu@gmail.com +79080569508,

Chokeberry thin layer convective drying process modeling and energy efficiency estimation

V Filipović[1,*], M Petković[2], J Filipović[3], N Miletić[2], I Đurović[2], J Radovanović[2] and A Lukyanov[4]

[1] University of Novi Sad, Faculty of Technology, Department of Chemical Engineering, Bul. cara Lazara 1, Novi Sad, Serbia
[2] University of Kragujevac, Faculty of Agronomy, Department of Food Technology, Cara Dušana 34, Čačak, Serbia
[3] University of Novi Sad, Institute for Food Technology, Research center for technology of plant based food products, Bul. cara Lazara 1, Novi Sad, Serbia
[4] Don State Technical University, square Gagarin 1, Rostov on Don, Russian Federation

*e-mail: vladaf@uns.ac.rs

Abstract. The effective moisture diffusivity (D_{eff}), the energy of activation (E_a) and total energy input (Q) are appropriate parameters for modeling and energy efficiency estimation of the thin layer dehydration process of chokeberry under the different air temperature (50, 55, 60, 65 and 70 °C). Increasing the dehydration temperature will statistically significantly increased the D_{eff} from 2.93×10^{-6} m^2 s^{-1} (50 °C) to 7.77×10^{-6} m^2 s^{-1} (70 °C), and decreased the E_i from 2594.77 kJ to 2474.35 kJ, respectively. As the drying temperature was increased and consecutive drying time has been reduced (from 37 hours to 23 hours), energy input statistically significantly decreased. Mathematical models showed a good correlation between calculated and experimental values and allowed good prediction of dehydration parameters, enhancing energy efficiency. The E_a for the experimental model of thin layer chokeberry dehydration was 45.37 kJ mol^{-1}.

1. Introduction

Food drying is a complex process because of the differential structure of products, with simultaneous heat and mass transfer [1]. In practice, a food dryer and effective models of drying are necessary for dehydration design, optimization, energy efficiency estimation, and process control. The drying parameters, such as drying temperature, pressure, air velocity, relative humidity, material thickness and shape, and the product retention time, must be controlled according to product type, feed, purpose, and method [2, 3, 4].

Thin-layer drying is used to estimate the drying time for several products (such as fruit berries) and also to generalize energy efficiency. Physical and thermal properties of fruit berries (chokeberry), such as the heat and mass transfer, and moisture diffusion, are essential for the dryer design. The diffusion process in solid materials during the dehydration process is very complicated. It may involve molecular diffusion, different flows (capillary, Knudsen, hydrodynamic), or surface diffusion [1]. All these phenomena are systematized as effective moisture diffusivity.

The goal of this research was to develop a mathematical model of the chokeberry thin layer convective drying process and to calculate and estimate process energy efficiency at the different process parameters.

2. Materials and methods

2.1. The theoretical background of the dehydration process

Convective drying was conducted in the dehydrator (Colossus CSS 5330 250W, PRC) at temperatures of 50, 60, and 70 °C at atmospheric pressure, to the constant weight. Uncrushed chokeberry samples were placed in a tray of 320-mm diameter in a thin layer with a mass load of 3 kg m^{-2}, and the convective drying process was operated at an air velocity of 0,25 m s^{-1} [5]. The convective drying kinetic was based on mass losses of chokeberry [6]. The moisture ratio (MR) is defined according to Eq. (1):

$$MR = \frac{M_t - M_e}{M_o - M_e} \tag{1}$$

M_t, M_o and M_e represent the moisture content achieved after convective drying time t, the initial moisture content, and the equilibrium moisture content, respectively. The value of equilibrium moisture content (M_e) usually is deficient and can be deleted from Eq. (1) without a significant change in the amount of *MR*.

The most widely used theoretical models are derived from Fick's second law of diffusion. The analytical solutions for berries (sphere, Eq. (1) and (2)) were given below [1]:

$$MR = A_1 \cdot \sum_{i=1}^{\infty} \frac{1}{J_o^2} \cdot e^{-\frac{J_0^2 \cdot D_{eff}}{A_S}} \tag{2}$$

$$A_1 = \frac{6}{\pi^2}; \quad A_2 = 4 \cdot r^2 \tag{3}$$

where, D_{eff} is the effective moisture diffusivity (m^2 s^{-1}), t is time (s), MR is the moisture ratio, J_0 is the roots of the Bessel function, A_1, A_2 are geometric constants, and r is the radius of the sphere.

Eq. 1 is derived for the constant values of D_{eff}, and for sufficiently long drying time:

$$ln(MR) = ln(a) - k \cdot t \tag{4}$$

$$k = -\frac{\pi^2 \cdot D_{eff}}{A_2} \tag{5}$$

The variation of $ln(MR)$ versus t is linear (Eq. (4)), and the slope is equal to drying constant (k) and the constant effective moisture diffusivity can be easily calculated (Eq. (5)).

The energy of activation (E_a) was calculated using an Arrhenius type equation (Eq. (6)) [7]:

$$D_{eff} = D_0 \cdot e^{-\frac{E_a}{R \cdot T}} \tag{6}$$

where E_a is the energy of activation (kJ mol^{-1}), R is the universal gas constant (8.3143 J mol^{-1} K^{-1}), T is absolute air temperature (K), and D_0 is the pre-exponential factor of the Arrhenius equation (m^2 s^{-1}). The activation energy was calculated from the slope of the Arrhenius plot $ln(De)$ versus $1/T$, given by Eq. (7):

$$n = -\frac{E_a}{R} \tag{7}$$

2.2. Energy input calculation

Energy input (Q) during chokeberry drying was calculated via a simple experiment where the mass of distilled water was poured in a full vessel, and its temperature was measured. Then vessel with water was placed in a preheated convective dryer at defined temperatures by the experimental plan. After 10 minutes, water loss, expressed in weight, is measured. Quantity of heat needed for increase of the temperature of water samples and evaporation of determined mass of water is calculated from the following equation [8]:

$$Q = c_p \cdot m_s \cdot (T_2 - T_1) + L \cdot m_i \tag{8}$$

where are:

Q – a quantity of heat needed for heating and evaporation of determined mass of water, (kJ)

c_p – specific heat capacity of the water, (kJ kg^{-1}°C)
m_s – sample mass, (kg)
T_2 – final sample temperature, (°C)
T_1 – start sample temperature, (°C)
L – latent heat of water evaporation, (kJ kg^{-1})
m_i – a mass of evaporated water from the sample, (kg).

The calculated quantity of heat for 10 minutes was extrapolated to experimentally measured drying time of chokeberry until a constant mass was achieved.

2.3. Mathematical models calculation

Based on experimental data of drying time and coefficient of diffusion, mathematical models of the dependence of drying time and coefficient of diffusion from drying temperature, as an independent variable, are formed:

$$Y_k = f_k(\textit{drying temperature}) \quad (9)$$

The second-order polynomial is used for the approximation of experimental data. Models of 3 responses (drying time, coefficient of diffusion, and energy input) in dependence of 1 independent variable (drying temperature) are developed:

$$Y_k = a_{k0} + a_{k1} \cdot X + a_{k2} \cdot X^2 \quad k = 1\text{–}3; \quad (10)$$

where are:
a_{k0-2} – regression coefficients;
Y – drying time (Y_1), coefficient of diffusion (Y_2), energy input (Y_3)
X – drying temperature.

The significance of the effect and interaction of individual factors, for every response, was determined by analysis of variance (ANOVA) and application of post-hoc Tukey HSD test. For ANOVA calculation, StatSoft Statistica ver.12.0. The software package was used, and for and second-order polynomial models' calculation Microsoft Excel ver. 2016 was used.

3. Results and Discussion

D_{eff} mainly varies with the product's temperature, the moisture content, air velocity, and the structure. It is in accordance with the thin layer assumption [9, 10]. Chokeberries were relatively homogeneous spheres with a diameter of 5.38±0.25 mm.

The values of drying times and the coefficient of diffusion at different drying temperatures are shown in Figure 1.

From the presented results, it can be seen that with the increase of drying temperature, a statistically significant decrease of drying duration occurred, until the constant mass of dried chokeberry is obtained. The longest drying duration was 37 hours, at the lowest drying temperature (50 °C), while the shortest drying duration of 23 hours was achieved at the highest drying temperature of 70 °C. The values of the coefficient of diffusion statistically significantly increased with the increase in drying temperature. These results are in a correlation to Mirzae et *al.* (2009) and Aghbashlo et *al.* (2008). It can be seen that there was no statistically significant difference between the coefficient of diffusion values at the two highest drying temperatures, which indicate reaching maximal values of the coefficient of diffusion at these temperatures. Minimal value of coefficient of diffusion was 2.93×10^{-6} $m^2 s^{-1}$ ($ln(MR) = -0.0997 \cdot t + 0.1709$, R^2=0.9868), at drying temperature of 50 °C, while maximal value of 7.77×10^{-6} $m^2 s^{-1}$, was at drying temperature of 70 °C ($ln(MR) = -0.2647 \cdot t + 0.5641$, R^2=0.9587). Increasing the dehydration time, the slope (k) was increased too.

The activation energy (E_a) was related to the minimum amount of energy that needs to be provided to initial moisture diffusion through the chokeberries and was about n = 45.37 kJ mol^{-1}. E_a varies depending on the drying parameters and ranges from 29.35–33.78 kJ mol^{-1} for apricot to 110.837–130.61 for kJ mol^{-1} particular types of berries [2, 12].

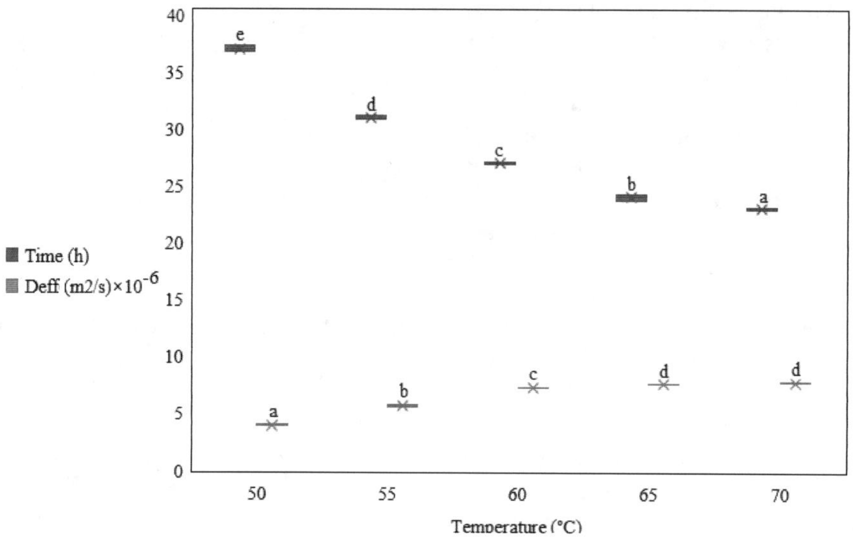

[a-d] Different letters in superscript between marked values of the same color, indicates on statistically significant difference, at the level of significance of $p<0.05$ (based on post-hoc Tukey HSD test)

Figure 1. Average values and standard deviations of drying time and coefficient of diffusion at different drying temperatures

In Table 1, values of Q during chokeberry drying at different drying times and temperatures are shown. From the presented result, it can be seen that with the drying temperature increase and consecutive drying time decrease, energy input statistically significantly decreased. The highest value of energy input was achieved at drying temperature and time of 50 °C and 37 h. In comparison, the lowest energy input was made at drying temperature and time 70 °C and 23 h, indicating that a more energy-efficient process was at the highest temperature.

Table 1. Average values and standard deviations of energy input during chokeberry drying at different drying time and temperature

T (°C)	t (h)	Q (kJ)
50	37	2594.77 ± 11.15^{b}
55	31	2578.28 ± 9.71^{bc}
60	27	2560.10 ± 12.34^{b}
65	24	2489.46 ± 14.06^{a}
70	23	2474.35 ± 15.74^{a}

[a-c] Different letters in superscript in the Table 1 indicate on the statistically significant difference between values, at a significance level of $p<0.05$

Mathematical modeling of the thin layer dehydration of fruits usually requires the statistical methods of regression and correlation analysis. Table 2 shows regression coefficients and coefficients of correlation of developed mathematical models of drying time and coefficient of diffusion from drying temperature in chokeberry convective drying process, in the form of second-order polynomial Eq. (3). Based on second-order polynomial Eq. (3) regression coefficients ($a_0 - a_2$), mathematical models, of drying time, and the coefficient of diffusion dependence from drying temperature in the chokeberry convective drying process can be formed. High R^2 values indicate the high correspondence of calculated values with experimental data [13].

Table 2. Mathematical models of drying time and coefficient of diffusion dependence from drying temperature in chokeberry convective drying process

		t	D_{eff}	Q
Regression coefficients	a_0	181.97	-74.358	2430.9
	a_1	-4.4714	2.4876	10.444
	a_2	0.0314	-0.0188	-0.142
R^2		0.999	0.993	0.941

4. Conclusions

From the presented results, it can be concluded that:

- By increasing temperature in the process of chokeberry drying, the duration of the process can be significantly shortened, reaching maximal values of the coefficient of diffusion
- The experimental results showed that at a high air temperature (70 °C), the air had better contact with the chokeberry surface, which results in more excellent absorption of moisture. The moisture gradient was increased, which was leading to an increase in the moisture diffusivity
- Developed mathematical models of drying time and coefficient of diffusion dependence from drying temperature in chokeberry convective drying process showed a good correlation between calculated and experimental values and allowed the right prediction of process parameters
- With increasing process temperature, energy input decreased, enhancing energy efficiency.

5. Acknowledgments

Presented results are obtained in a project supported by the Ministry of Education, Science and Technological Development of the Republic of Serbia, grant number: 451-03-68/2020-14/ 200134.

References

[1] Erbay Z and Icier F 2010 *Crit. Rev. Food Sci. Nutr.* **50** 441-464
[2] Mirzae E, Rafiee S, Keyhani A and Emam-Djomeh Z 2009 *Res. Agr. Ang.* **55** 114-120
[3] Anawe PA L and Adewale FJ 2018 Data on exergy and exergy analyses of drying process of onion in a batch dryer. *Data Brief.* **21** 1784
[4] Argo BD and Ubaidillah U 2020 *J. Mech. Sci. Technol.* **34** 435-442
[5] Petković M, Đurović I, Miletić N and Radovanović J 2019 *Period. Polytech. Chem. Eng.* **63** 600
[6] Szychowski PJ, Lech K, Sendra-Nadal E, Hernández F, Figiel A, Wojdyło A and Carbonell-Barrachina ÁA 2018 *Food Bioprocess Tech.* **255** 157-164
[7] Lopez A, Iguaz A, Esnoz A and Virseda P 2000 *Dry. Technol.* **18** 995-1006
[8] Toledo RT 2007 *Fundamentals of Food Process Engineering* (University of Georgia, Athens, Georgia, USA)
[9] Ece MC and Cihan A 1993 *T. Asabe.* **36** 837-840
[10] Akpinar E, Midilli A and Bicer Y 2003 *Energ. Convers. Manage.* **44** 1689-1705
[11] Aghbashlo M, Kianmehr MH and Samimi-Akhijahani H 2008 *Energ. Convers. Manage.* **49** 2865
[12] Garau MC, Simal S, Femenia A and Rossello C 2006 *J. Food Eng.* **75** 288
[13] Arsenović M, Pezo L, Stanković S, Radojević Z 2015 *Appl. Clay Sci.* **115** 108-114

IOP Conf. Series: Materials Science and Engineering **900** (2020) 012002 doi:10.1088/1757-899X/900/1/012002

Synthesis of a nonlinear control law with efficiency energy for the self-balancing two wheeled vehicle

C X Nguyen[1,*] , A D Lukianov[2], T D Pham[3] and A D Nguyen[4]

[1]Department of Automation and Computing Techniques, Le Quy Don Technical University, Vietnam

[2]Department of Automation of production processes, Don State Technical University, Rostov-on-Don, Russia

[3]Department of Technology, equipment and aerospace, Le Quy Don Technical University, Vietnam

[4]Academy of science and technology, Vietnam

*e-mail: nguyenxuanchiem83@gmail.com

Abstract. In the paper, we present a method of synthesizing the controller with optimum energy for two-wheeled self-balancing vehicles, based on the multi-purpose consumption function. The authors select the multi-purpose consumption function based on the quasi-optimality and minimum of energy. The synthesis of control laws based on the analytical design of aggregated controllers (ADAR) with manifold of quasi-optimality. The results of the proposed method are simulated and compared to other methods.

1. Introduction

A two-wheeled self-balancing vehicle has two wheels with two the same axles, built into a commercial product derived from the original idea of the Segway company. The principle of controlling a two-wheeled self-balancing vehicle is to combine the position control of the person on the vehicle and the automatic balance controller for the vehicle. When you want the car to move straight, the driver just needs to stand straight and lean forward slightly. If you want to increase the speed of the vehicle's movement, the person facing forward more. To stop the driver, just stand upright so that it is perpendicular to the platform, this time the car will automatically stop and still be balanced on the vehicle (fig. 1) [1].

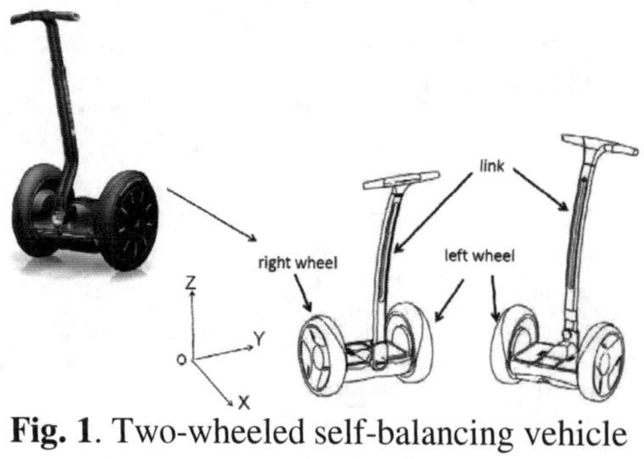

Fig. 1. Two-wheeled self-balancing vehicle

This is a means of transportation that is used more and more widely with many advantages: compact, light, takes up little area when moving, does not cause environmental pollution, is easy to control, flexible movement in many places, easy to transport, easy to disassemble.

Two-wheeled self-balancing vehicle works based on the inverted pendulum. This is the standard model for developing control algorithms. There has been a lot of control algorithms for this object. In [2, 3] using traditional PID controllers are presented with various methods of determining controller parameters. The studies [4, 5] built the controller according to LQR method but did not consider the energy consumption through control signals. The study [6] proposed a method to use Type-2 fuzzy controller to simplify in building control laws. The studies [7, 8] proposed a control law with quasi-optimality, ensuring system quality, but is not energy efficient. A critical issue with vehicles is energy efficiency when operating. In this paper, the authors propose a method to synthesize the controller with energy efficiency base a synergistic approach by ADAR method to the target function and manifold with quasi-optimality.

2 The mathematical model of the self-balancing two wheeled vehicle

We consider the well known the self-balancing two wheeled vehicle [1,] which is schematically depicted in Fig. 2.

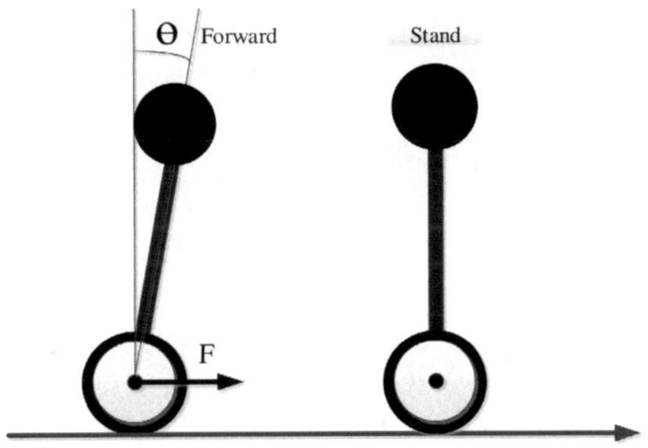

Fig. 2. The self-balancing two wheeled vehicle

The linear mathematical model of two-wheeled self-balancing vehicle at the equilibrium point is described in equations (1) which was presented in [7]:

$$\begin{bmatrix}\dot{x}\\ \ddot{x}\\ \dot{\theta}\\ \ddot{\theta}\end{bmatrix}=\begin{bmatrix}0 & 1 & 0 & 0\\ 0 & \dfrac{k_e k_e(M_p lr-I_p-M_p l^2)}{Rr^2\alpha} & \dfrac{M_p l^2}{\alpha} & 0\\ 0 & 0 & 0 & 1\\ 0 & \dfrac{2k_e k_e(r\beta-M_p l)}{Rr^2\alpha} & \dfrac{M_p^2 g l^2\beta}{\alpha} & 0\end{bmatrix}\begin{bmatrix}x\\ \dot{x}\\ \theta\\ \dot{\theta}\end{bmatrix}+\begin{bmatrix}0\\ \dfrac{2k_m(I_p+M_p l^2-M_p lr)}{Rr\alpha}\\ 0\\ \dfrac{2k_m(M_p l-r\beta)}{Rr\alpha}\end{bmatrix}V_a \tag{1}$$

where $\beta=(2M_\omega+\frac{2I_\omega}{r^2}+M_p)$; $\alpha=(I_p\beta+2M_p l^2(M_\omega+\frac{I_\omega}{r^2}))$; x - displacement (m); $\dot{x}$ - velocity of vehicle (m/s); $\ddot{x}$ - acceleration of vehicle (m/s2); θ - angular displacement (rad); $\dot{\theta}$- angular velocity (rad/s);$\ddot{\theta}$- angular acceleration (rad/s2);Va- armature voltage (V); km - motor's torque constant (Nm/A); ke - back EMF constant (Vs/rad.); R - motor resistance (Ω); l - distance from wheel axle to center of chassis (m); g- gravitational constant (m/s2); Mp- mass of the chassis (kg); r- radius of wheels (m); Ip- moment of inertia of the chassis (Nm); Iω - moment of inertia of the wheels (Nm); Mω - mass of the wheels (kg).

In this article, we need to design a angular controller for your vehicle. The control system block diagram for two-wheeled self-balancing vehicle is as shown in Figure 3.

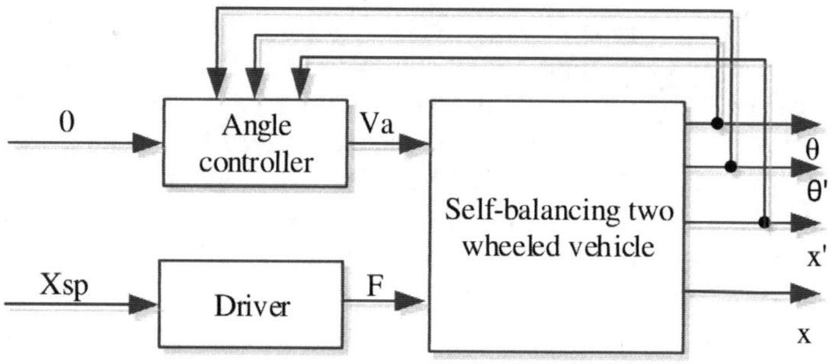

Fig. 3. The control system block diagram for two-wheeled self-balancing vehicle

From the control system block diagram, the driver will apply the force to the car so that the angle deviates from the zero balance point forward. Therefore, in this paper when designing the angle controller, from the system of equation (1), we get the state equation according to the angle ($x_1 = \theta, x_2 = \dot{\theta}$) of two-wheeled self-balancing vehicles as follows:

$$\begin{bmatrix} \dot{x}_1 \\ \dot{x}_2 \end{bmatrix} = \begin{bmatrix} 0 & 1 \\ \dfrac{M_p{}^2 g l^2 \beta}{\alpha} & 0 \end{bmatrix} \begin{bmatrix} x_1 \\ x_2 \end{bmatrix} + \begin{bmatrix} 0 \\ \dfrac{2k_m (M_p l - r\beta)}{Rr\alpha} \end{bmatrix} V_a + \begin{bmatrix} 0 \\ \dfrac{2k_m k_e (r\beta - M_p l)}{Rr^2\alpha} \end{bmatrix} \dot{x} \quad (2)$$

3 Synthesis of energy efficient angular controller based on synergistic method ADAR

a. Synergistic approach to the analytical design of aggregated regulators.

The analytical design of aggregated controllers (ADAR) proposed in [9, 11] in the sense of ensuring asymptotic stability and given dynamic properties of transients is based on the introduction of functionals of the form:

$$J = \int_0^\infty F(\psi, \dot{\psi}) dt \quad (3)$$

where $F(\psi,\dot{\psi})$ – continuously differentiable function with all arguments; $\psi(x_1, \ldots x_n)$ – aggregated macro-variable, representing some arbitrary or piecewise of differentiable continuous function of phase coordinates $(x_1, \ldots, x_n)$, moreover $\psi(0, \ldots, 0) = 0$.

The essence of the synergetic approach is to construct an accompanying function in such a way as to determine in the phase space of the non-linear object under consideration a certain set of points to which all close motion paths are attracted. Such a set of points is usually called an attractor [10]. The above actually corresponds to the basic principle of synergetics - the principle of submission, on which the theory of self-organization of nonlinear dynamic systems is built [9].

In particular, for a non-linear SISO object, with scalar control, it is advisable to choose the following quadratic form as an integrand in (4):

$$F(\psi,\dot{\psi}) = m^2\varphi^2(\psi) + c^2\dot{\psi}^2(t) \quad (4)$$

with the function $\varphi(\psi)$ must satisfy the following conditions:

1. Single-valuedness, continuity and differentiability for all values;
2. $\varphi(0) = 0$;
3. $\varphi(\psi)\psi > 0$ at any $\varphi \neq 0$.

b. Synthesis of angle stability control laws for two-wheeled self-balancing vehicle that are energy efficient

The energy-saving for autonomous vehicles is current issue but there has not been much research. In order to synthesize control laws with energy optimization, studies have shown the target function of energy optimality. The requirement of controller is energy efficient and the system quality is the same as the system with the quasi-time optimal control law. The authors propose a method of synthesizing control laws based on synergistic method ADAR.

Select macro-variables with quasi-optimality $\psi = x_2 + \nu \frac{x_1}{\sqrt{x_1^2 + \varepsilon_1^2}}$. Select the target function for the equations (2) of the form (5):

$$J = \int_0^{\infty} (\varphi^2(\psi) + T^2 \dot{\psi}^2(t))dt \tag{5}$$

where

$$\varphi(\psi) = \frac{\psi}{\sqrt{\psi^2 + \varepsilon_2^2}},\ \dot{\psi} = \frac{2k_m(M_p l - r\beta)}{Rr\alpha} V_a + \frac{M_p^2 g l^2 \beta}{\alpha} x_1 + \frac{2k_m k_e (r\beta - M_p l)}{Rr^2 \alpha} y + \frac{\nu \varepsilon_1^2}{(x_1^2 + \varepsilon_1^2)^{1.5}};$$

From the target function as chosen when away from the equilibrium point $\varphi^2(\psi) \to 1$, the target function becomes $J = \int_0^{\infty} (1 + T^2 \dot{\psi}^2(t))dt$. Clearly, the target function contains the optimal component in terms of energy and the quasi-time optimization.

The law of synthetic control must satisfy the functional equation (6) [9,11]:

$$\varphi(\psi) + T\dot{\psi}(t) = 0 \tag{6}$$

Solving equation (6) and receiving the control law in analytic form has the following form [7]:

$$V_a = -\frac{Rr\alpha}{2k_m(M_p l - r\beta)} \left(\frac{M_p^2 g l^2 \beta}{\alpha} x_1 + \frac{2k_m k_e (r\beta - M_p l)}{Rr^2 \alpha} \dot{x} + \frac{\nu \varepsilon^2}{(x_1^2 + \varepsilon^2)^{1.5}} + \frac{1}{T} \varphi(\psi) \right) \tag{7}$$

4 Simulating results of the angle stability control law of two-wheeled self-balancing vehicle

The parameters of mathematical model (1) is determined by real model as follows [1]: km = 0.022(Nm/A), ke = 0.4(Vs/rad), R= 1.7(Ω), l = 0.05 (m), g = 9.81(m/s^2), Mp=1 (kg), r= 0.0325(m), Ip=0.0012(kg.m^2), Iω=0.000016(kg.m^2), Mω=0.03(kg). The value of the controller parameters include v=1, ε1=ε2=1, T=0.0025. The simulation was performed on Simulink-Matlab software with the initial condition of the angle of vehicle x1=0.6 (rad), x2=0 (rad/s) with the purpose of bringing the vehicle to the balance position (x1=0, x2=0) with the quasi-optimality and energy-saving. The simulation results were performed with two controllers shown in fig. 4 and fig. 5: Quasi-time optimal controller (KOB) and proposed controller (SI-ADAR). In particular, fig. 4 (a) is the angle response of vehicle, fig. 4 (b) is the velocity response of vehicle, fig.5 (a) is the phase plane of the system. The results show that the system response to both controllers is almost identical, which means that the proposed controller retains the quasi-optimal characteristics of the KOB controller. In fig. 5 (b) is the control signal given by KOB controllers and the proposed method, the results show that the value of the proposed controller is more energy efficient.

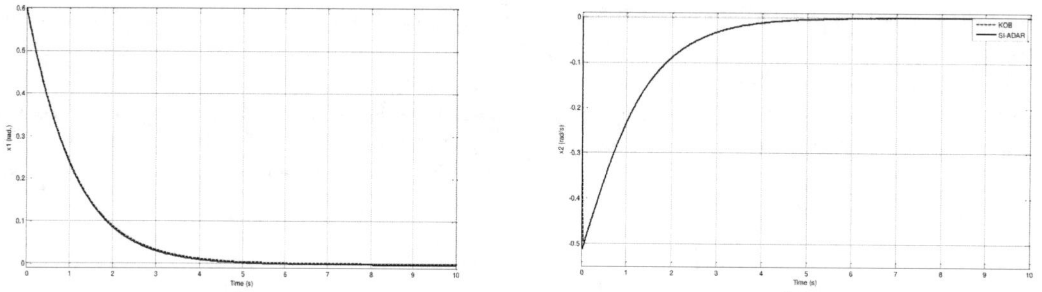

a) b)

Figure 4. a) The angle response of vehicle, b) The velocity response of vehicle

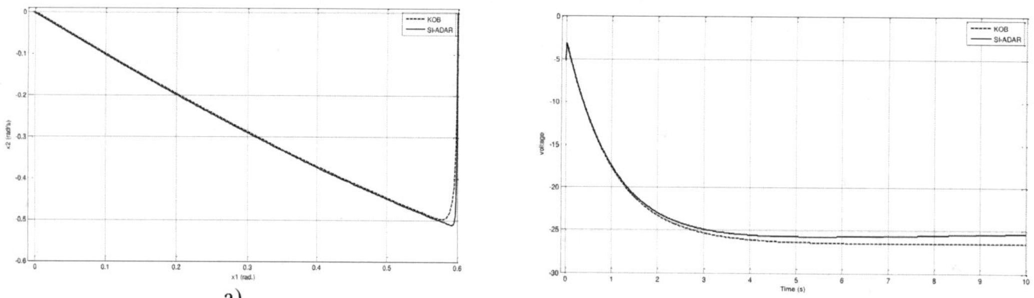

a) b)

Figure 5. a) The phase space of system, b) The control signal

5 Conclusion

Combining the ADAR method with the determined manifold method with quasi-optimality and the target function with energy efficient, the authors have built a control law that is both quasi-optimality and energy efficient. Comparisons with the quasi-time optimal controller show the effectiveness of the proposed method. The control signal power of the proposed method is smaller. Besides, we also see that with the choice of such a target function, when near the control signal equilibrium point there is no energy optimization. Therefore, in future studies, the authors will present methods to overcome these disadvantages.

References

[1] Soufiane Haddout. Nonlinear reduced dynamics modelling and simulation of two-wheeled self-balancing mobile robot: SEGWAY system. Systems science & control engineering: an open access journal, 2017 https://doi.org/10.1080/21642583.2017.1413436.

[2] Eldhose P Kunnel, Dijoy Johny, Anuja Jacob, Elizabeth Paul. Controlling of Two Wheeled Self Balancing Robot using PID. International Journal of Advanced Research in Electrical, Electronics and Instrumentation Engineering. Vol. 7, Issue 3, March 2018.

[3] G. M. T. Nguyen, H. N. Duong, and H. P. Nguyen, “A PID backstepping controller for two-wheeled self-balancing robot,” in Proceedings fo the2010 IFOST, 2010.

[4] Jian Fang. The LQR Controller Design of Two-Wheeled Self-Balancing Robot Based on the Particle Swarm Optimization Algorithm. Hindawi Publishing Corporation Mathematical Problems in Engineering Volume 2014, Article ID 729095, 6 pages http://dx.doi.org/10.1155/2014/729095.

[5] J. Zhao and X. Ruan, “Te LQR control and design of dual-wheel upright self-balance robot,” in Proceedings of the 7th World Congress on Intelligent Control and Automation (WCICA ’08), pp. 4859–4863, June 2008.

[6] Nguyen X.C., Phan N.H., Truong D.K., Hoang D.L., Tran C.P.,Do N.S. (2018) *Inzhenernyy vestnik Dona*, 3. ivdon.ru/ru/magazine/archive/n3y2018/5220.

[7] Hai N. Phan, Chiem X. Nguyen, “Building embedded quasi-time-optimal controller for twowheeled self-balancing robot,” MATEC Web Conf.Volume 132, 2017 XIII International Scientific-Technical Conference“Dynamic of Technical Systems” (DTS-2017).

[8] Nguyen X. C.,Phan H.N.. Design controler of the quasi time optimization approach forstabilizing and trajectory tracking of inverted pendulum. MATEC Web Conf.Volume 226, 2018 XIV International ScientiÒc-Technical Conference “Dynamic of Technical Systems” (DTS-2018).

[9] Колесников А.А. Синергетическая теория управления – Таганрог: ТРТУ, Энергоатомиздат, 1994. – 344 с.

[10] А.А. Стенин, В.П. Пасько, Е.Ю. Мелкумян, М.А. Солдатова. Анализ проблемы синтеза систем управления нелинейными динамическими объектами. Міжвідомчий науково-технічний збірник «Адаптивні системи автоматичного управління», 2015, № 2(27).

[11] Колесников А.А. Синергетика и проблемы теории управления: сб. научн. тр. Под ред. А.А. Колесникова. М.: ФИЗМАТЛИТ, 2004.

The relationship between the vibration energy of cutting and the wear resistance of the cutting wedge

V P Lapshin*, V V Khristoforova

Don State Technical University, Rostov-on-Don, Russian Federation

*e-mail: Lapshin1917@yandex.ru

Abstract. This work is an attempt to assess the impact of tool vibrations during cutting on the stability of the cutting wedge. In this paper, a numerical experiment is conducted to model the wear process of a cutting wedge based on the use of the Volterra operator of the second kind. The article presents the results of modeling the Volterra operator for various variants of vibration cutting processes that differ from each other in the amount of energy of the observed vibration signal. The results of experiments showed that the stability of the cutting wedge decreases with increasing vibration energy of the tool. This is due to the fact that when the vibration dynamics of cutting increases, the actual path traversed by the tool during processing increases, which increases the total work performed when processing a given part.

1. Introduction

In addition to the parameters that determine the state of the cutting process, the parameters that characterize the quality of manufactured parts are also changed. It is known that during the processing process, the geometry of the formed part and its maximum and minimum geometric properties (undulation, roughness, sub-roughness) change. All these changes are associated with the evolution of the properties of the dynamic cutting system. The most significant wear factor that affects the quality of the treated surface is wear on the back face, this type of tool wear is a well-experimentally studied phenomenon, which is referred to in the literature as: - h3 . The basis of the approach to modeling such dependencies is first presented in [1-2], which deals with General issues of using Volterra integral operators for modeling tool wear. The calculation of wear on the back face based on a set of discrete arrays of values of operation and power of irreversible transformations was not carried out in these works. When representing wear in state coordinates, we rely on the relationship between wear and work by the power of irreversible transformations of the input energy in the cutting zone. Fundamentally, the change in wear depending on irreversible transformations is revealed in the works of A. L. Bershadsky, B. I. Kostetsky, A. A. Ryzhkin, and others [3-5]. The reason for evolutionary transformations is related to the power and operation of cutting forces, that is, the energy of irreversible transformations in the processing zone. Thus, all parameters that characterize the cutting process and the quality of manufacturing parts are not only interrelated, but also have the same nature of change - it is the trajectory of the power of irreversible transformations for perfect work. The issue of experimental estimation of the cutting wedge wear is also of interest. more accurate is the statement about the convergence of numerical solutions considered when modeling wear using the Volterra operator with the results of numerous experiments described in [6-9]. One of the features identified during these experiments was that the wear rate is associated with the deformation dynamics of cutting. The purpose of this paper is to evaluate the effect of the deformation dynamics of cutting on the wear rate when modeling wear by the Volterra operator

2. Description of the mathematical model and numerical experiment

To model evolutionary changes in this case, it is necessary to use Volterra integral operators of the second kind, which have the following structure [1]:

$$h_3 = k\int_0^A w(t-\xi)\cdot N(\xi)d\xi \tag{1}$$

where $w(t-\xi)$ - core of the integral operator, $N(\xi)$ - phase trajectory of the power of irreversible transformations on perfect operation, A- operation of cutting forces. However, as follows from the structure of the operator, wear depends on the fluidity of irreversible transformations and its background, which is taken into account by the kernel of the integral operator (see expression 1). The process of tool wear is always bi-directional, one of the directions of which is focused on the addition of the cutting wedge to the cutting process, as a result of which we observe, described in the previous sections, the process of forming the contact area of the tool on the back face, through which the temperature field and force reaction are stabilized. The second direction is associated with an increase in the degradation features of the wear process, which subsequently lead to a significant change in the cutting properties and properties of contact of the tool and a sharp increase in cutting forces and associated vibrations. Based on these arguments, it is convenient to consider the kernel of an integral operator as the sum of two kernels, as it is represented in the expression (2):

$$h_3 = \int_0^A (\beta_1 e^{\alpha_1(\xi-A)} + \beta_2 e^{\alpha_2(A-\xi)})N(\xi)d\xi\,, \tag{2}$$

where $(\beta_1 e^{\alpha_1(\xi-A)} + \beta_2 e^{\alpha_2(A-\xi)})$ - the sum of the kernels of the integral operator, while $\beta_1 e^{\alpha_1(\xi-A)}$ - the core that determines the tool burn-in trajectory, and $\beta_2 e^{\alpha_2(A-\xi)}$ - the core that determines the path of degraded wear, $\beta_1, \beta_2, \alpha_1, \alpha_2$ - parameters to be identified.

The power of irreversible transformations is determined based on the expression (3), and the work can be defined as the following integral - $A = \int_0^t N(t)dt$.

For the case when N=N0=const, (2) is conveniently represented as:

$$h_3 = N_0\int_0^A (e^{\alpha_1(\xi-A)} + e^{\alpha_2(A-\xi)})d\xi\,. \tag{3}$$

The solution for (3) is as follows:

$$h_3 = N_0\frac{1}{\alpha_1}(1-e^{-\alpha_1 A}) + N_0\frac{1}{\alpha_2}(e^{\alpha_2 A}-1)\,. \tag{4}$$

If N(t) is not equal to N0=const, then we can assume that the measured or calculated discrete power values N0, N1,...., Nn-1 due to the smallness of the value ΔA=Ak-Ak-1, close to constant or constant, we can get the following approximate discrete sum describing the integral operator (3):

$$\begin{aligned} h_3 = {} & \frac{\beta_1}{\alpha_1}[-N_0 e^{-\alpha_1 A} - (N_1-N_0)e^{\alpha_1(A_1-A)} - (N_2-N_1)e^{\alpha_1(A_2-A)} - \ldots - (N_{n-1}-N_{n-2})e^{\alpha_1(A_{n-1}-A)} + N_{n-1}] + \\ & + \frac{\beta_2}{\alpha_2}[N_0 e^{\alpha_2 A} + (N_1-N_0)e^{-\alpha_2(A-A_1)} + (N_2-N_1)e^{-\alpha_2(A-A_2)} + \ldots + (N_{n-1}-N_{n-2})e^{-\alpha_2(A-A_{n-1})} - N_{n-1}] \end{aligned}\,. \tag{5}$$

Expression (5) can be modeled using the Matlab/Simulink package of mathematical programs. The simulation results for the case of a constant cutting force are shown in figure 1.

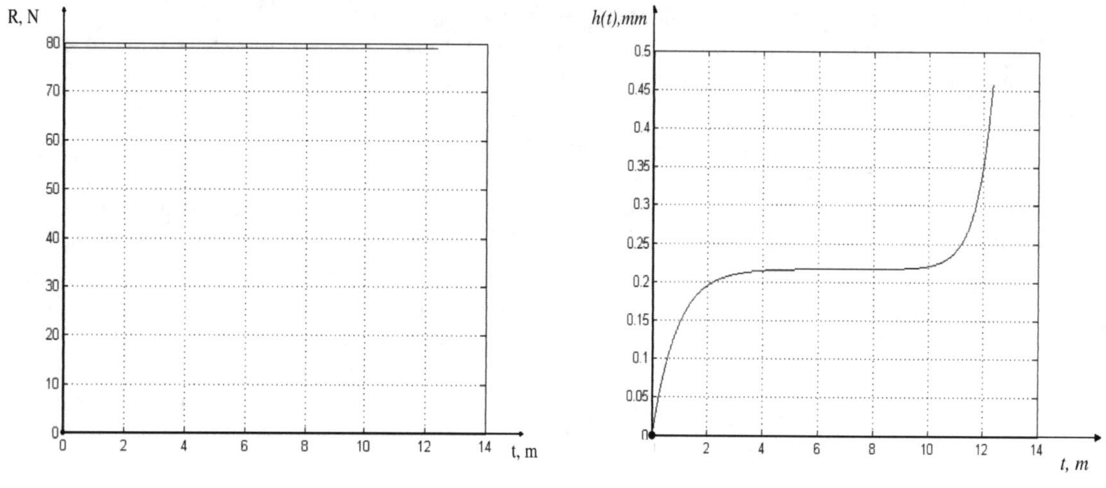

a) b)

Figure 1. Simulation results: a) constant cutting force graph, b) wear curve

As can be seen from figure 1, the wear curve obtained for this case of modeling the cutting control system is similar to the curve, close to the curves that we determined experimentally [6-9], but here the area of stabilization of the wear characteristic is more pronounced, as well as the time of critical wear increase is somewhat shorter. This was done specifically to clarify the results of subsequent modeling, the critical wear increase time is reduced due to the increased force reaction (see figure 2 a)), and the stabilization section is necessary for subsequent correction of the cutting system properties in this section. For subsequent modeling, it is of interest to consider the process of burn-in described by the first part of the discrete equation (5) and the increase in the process of degradation of the cutting wedge of the tool, the second part of the discrete sum represented in the expression (5). Figure 2 (a) shows a variant of modeling the curve that reflects the tool's running-in, and figure 2 (b) shows a variant of the curve that reflects the properties of increasing degradation of the tool during wear.

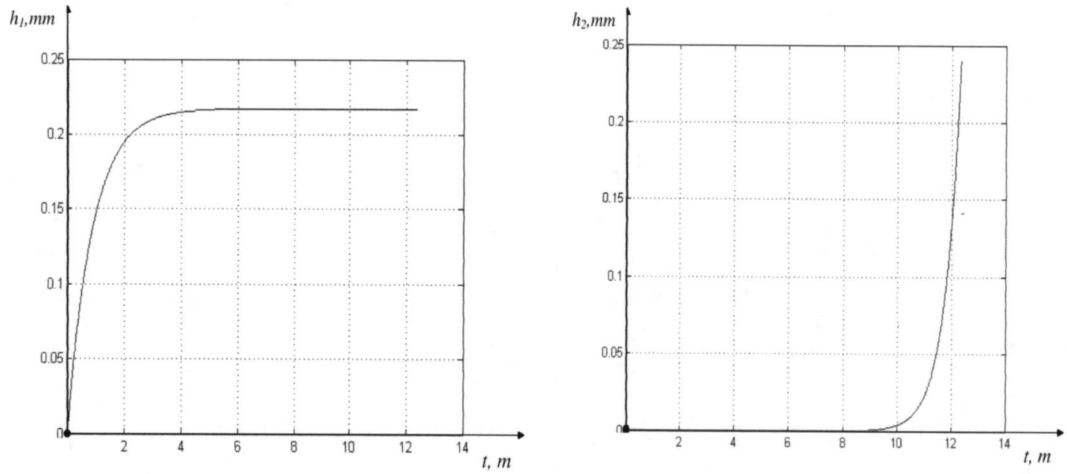

a) b)

Figure 2. Modeling: a) run-in, b) degradation

As you can see from figure 2 (a), the tool burn-in time is about 4 minutes, which basically corresponds to the wear schedule shown in figure 1, then this characteristic is stabilized and does not change until the end of the processing process. The value of 0.22 mm, accepted by us in this figure as a sidewall of stabilization of the burn-in process, is taken from the analysis of experimental results [6-9], where the effect of stabilization of the wear characteristic was observed in the vicinity of the nominal value. As can be seen from figure 2 (b), the tool degradation process begins to have a significant impact on the wear curve only after 9 minutes of processing. The curve reflecting the process of degradation of the cutting wedge is different from zero up to this minute, but its increase here is extremely small, as well as its influence on the entire wear curve up to 10 seconds of the cutting process is not great.

3. Results of the experiment and discussion

The only question of interest is whether tool vibrations will be taken into account in the run-in and degradation curves of the cutting wedge when modeling wear using the Volterra operator. To answer this question, we will model the Volterra operator with different types of vibration signals, which we will evaluate comparatively using the modulus of the vibration signal. This is due to the fact that direct integration of a complex cutting speed signal that has a periodic part in its composition will lead to the fact that the integral of the periodic component for the observation period will be equal to zero. For numerical estimation of the instrument vibration signal we used the following integral indicator:

$$VA = \sqrt{\left(\int_0^{T_v} \frac{dx(t)}{dt}^2 dt\right)}, \tag{6}$$

where $x(t)$ - the vibration signal itself, T_v - the period of observation of the signal.

To simplify our reasoning, we assume that we are talking about the scalar case of cutting, in which the cutting force is constant, and the speed of the relative movement of the tool along the workpiece is the sum of the cutting speed and the deformation vibrations of the tool. This allows us to describe the power of irreversible transformations as the product of the cutting force (F) and the sum of the cutting speed (V) and the speed of deformation movements of the tool (dx/dt), or as an expression:

$$M(t) = F(V + \frac{dx(t)}{dt}). \tag{7}$$

In this case it .is convenient to model the cutting force using the following expression:

$$A(t) = F(Vt + VA). \tag{8}$$

when modeling, we will use the initial data obtained earlier in experiments on a lathe [8-9], where the cutting speed V= 2620 mm/sec., the cutting force F=45 N., the amplitude of the tool's oscillation speed in the cutting direction was =50-200 mm/sec., the oscillation frequency was about 2000 Hz. Let's consider two extreme cases, in the first case we will take two extreme cases, the case with an oscillation amplitude of 50 mm / sec and the case of oscillations of 200 mm / sec. The simulation results are shown in figure 3.

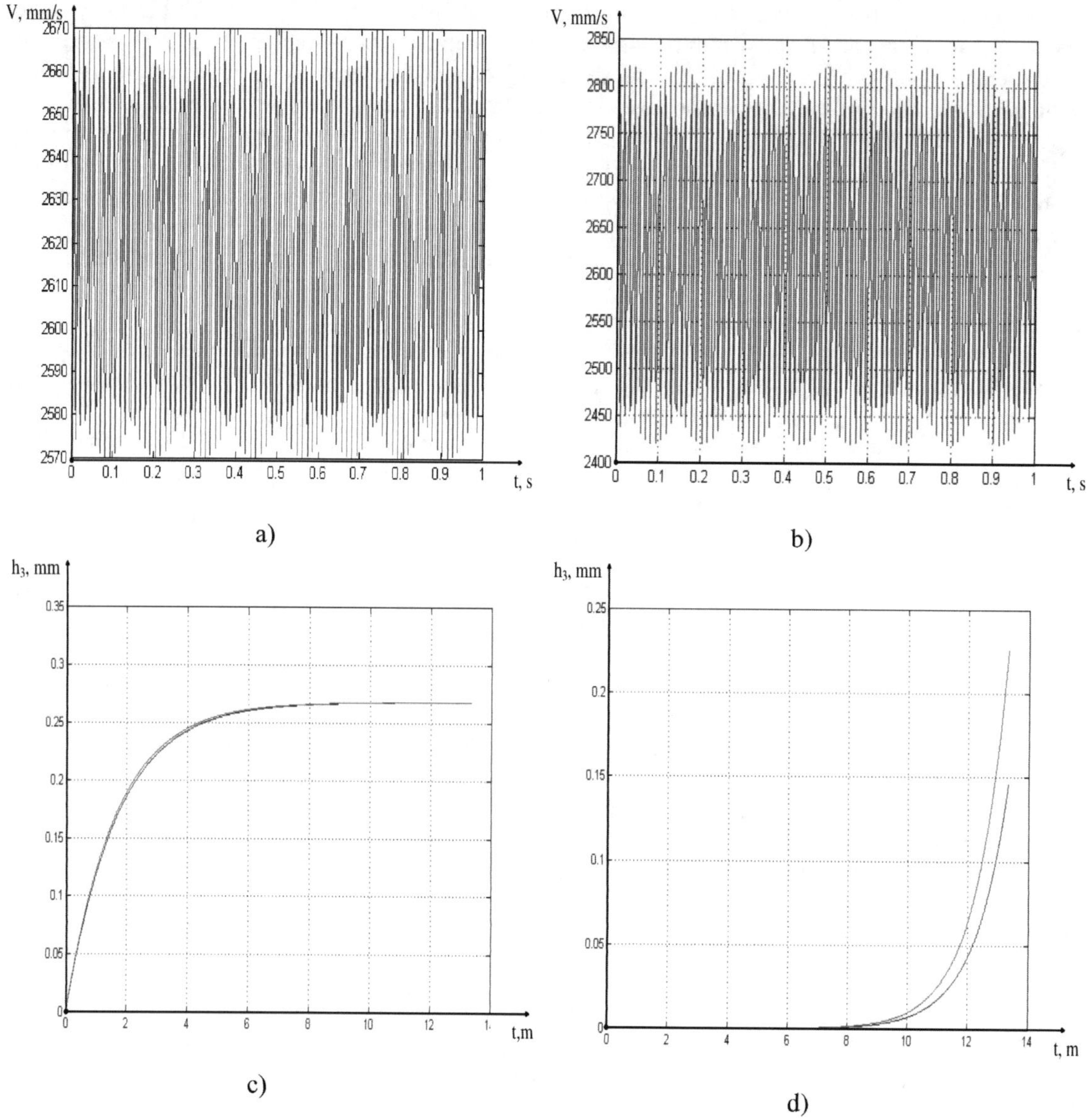

Figure 3. Simulation results: a)Speed signal for the case of 50 mm/s amplitude, b) speed Signal for the case of 200 mm/s amplitude, c) burn-in graphs, d) cutting wedge degradation graph.

As can be seen from figure 3, when the energy of the vibration signal reflecting the dynamics of the deformation movement of the tool increases by 4 times from 50 mm/sec in figure 3 (a) to 200 mm/sec in figure 3 (b), changes in the curve reflecting the tool wear are manifested only in part of the degradation characteristic (see figure 3 (d)) and almost do not appear in the run-in curve (see figure 3 (c)). Note that both characteristics, for example, show a characteristic associated with a higher amplitude variation of the instrument's vibrations.

Conclusions

Studies have shown that the use of a quadratic estimation of tool vibrations allows us to Refine the calculation of the wear curve of the cutting wedge, and the higher the amplitude or in other words, the energy of such a vibration movement, the less wear resistance of the cutting wedge.

IOP Conf. Series: Materials Science and Engineering **900** (2020) 012003 doi:10.1088/1757-899X/900/1/012003

References

[1] Zakovorotny V. L., Lapshin V. P., Babenko T. S. 2018. *Russian Engineering Research*. Vol. 38. P. 707-708.

[2] Zakovorotny V. L., Gvindjiliya V. E. 2019. *Russian Engineering Research*. Vol. 39. P. 423-430.

[3] Ryzhkin A.A. Synergetics of wear of tool materials during blade processing. Rostov-on-don: DSTU. 2019. 289 p. (in Russian).

[4] Kostetsky B.I., Bershadsky L.I., Chukreev E.N. 1970. *Reports of the USSR Academy of Sciences*. Vol. 191. H. 1339-1342 (in Russian).

[5] Postnov V.V., Shafikov A.A. 2012 *Vestnik UGATU*. Vol. 11, **2(29)**. P. 139-146 (in Russian).

[6] Lapshin V., Moiseev D., Minakov V. 2019. *AIP Conference Proceedings*. Vol. 2188.

[7] Zakovorotny V. L. et al. 2019. *E3S Web of Conferences*. EDP Sciences. Vol. 104.

[8] Lapshin V. P. et al. 2020. *Russian Engineering Research*. Vol. 40. P. 259-265.

[9] Lapshin V. P., Babenko T. S., Moiseev D. V. 2018. *International Conference on Industrial Engineering*. P. 853-859.

IOP Conf. Series: Materials Science and Engineering **900** (2020) 012004 doi:10.1088/1757-899X/900/1/012004

Method for diagnosing the state of bearings of spindle units by the energy of the vibration signal

V P Lapshin*, I A Turkin, Ya D Koba

Don State Technical University, Rostov-on-Don, Russian Federation

*e-mail: Lapshin1917@yandex.ru

Abstract. The work considers improved approach to assessing the serviceability of bearing assemblies of spindle groups of metal-cutting machines. As a basic malfunction, we looking a variant of the defect of the outer ring of the rolling bearing separator. For this task the calculated values of the vibration frequencies were obtained, and a full-scale experiment was conducted with the evaluation of the vibration signal on a large spectrum of frequencies. The results of the experiment showed that it is possible to assess the degree of bearing damage both from the spectrum of signal amplitudes and from the rms value of the vibration signal (RMS). However, studies have shown that the most beneficial, from the point of view of diagnostics, is to consider RMS not for the entire vibration signal, but only in the frequency range corresponding to the diagnosed bearing malfunction.

1. Introduction

The problem of analyzing the state of the main load-bearing systems of metal-cutting machines is extremely urgent today from a scientific, scientific and practical point of view. A lot of research is devoted to the assessment of the influence of vibrations on the state of metal processing by cutting [1-2]. Some studies consider the question of assessing vibrational activity from the point of view of the influence of temperature on it in the zone of vibration formation [3]. A number of works consider the effect of tool wear in metal-cutting equipment on the vibrational dynamics of the cutting process [4].
Most modern Russian researchers note the importance of controlling vibrational dynamics when processing metals on metal-cutting machines [5-7]. Here, authors lookin the diagnosics of processing processes during cutting, both during turning [5] and during milling [6], issues of general diagnostics of processes through analysis of a vibration signal are considered in [7]. On the whole, all these works show the ever-increasing role of vibration signal analysis as an information channel illustrating the processes occurring in the equipment under study.

2. Conditions for conducting experiments and mathematical methods used for analisis

When conducting the experiment, we must proceed from the fact that it is necessary to eliminate defects in the outer ring of the separator. An example of such a damage is shown in figure 1.

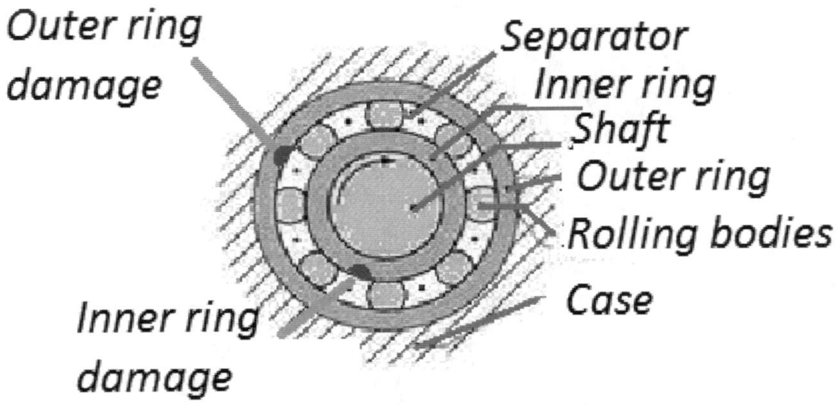

Figure 1. Conditional condition of defects

In the example presented in Figure 1, we are interested in the case of a defect in the outer ring of the bearing separator. For this, we used the SKF 6207-2z bearing, which is of Swedish origin, but today it is produced in our country. In order to diagnose a malfunction in this bearing in the form of a defect in the outer ring of the separator, we deformed him by hard blow. For the experiment, a specialized balancing machine was used, on which sensors for measuring vibration acceleration, manufactured by Vibropribor, DVST, were installed in the bearing mountings. The appearance of the experimental setup with installed sensors is shown in Figure 2.

Figure 2. Experimental stand

As can be seen from Figure 2, the sensors are mounted directly on the bearings, an engine belt is used to drive the rotor, and an optical speed sensor was used to measure the speed. To assess the influence of the dynamics of the machine associated with cutting, in which the cutting force acts on the spindle, we introduced an imbalance on the rotor, as shown in Figure 3.

Figure 3. Experimental stand with imbalance on the rotor

Let us consider that the introduction of an imbalance in the rotor will allow to simulate the cutting process, as it happens on the real spindle of the metal cutter.
Frequency of the beats connected with an imbalance will be connected with the frequency of rotation of a rotor which in an experiment we will establish in 2000 revolutions per minute, or in 33.33 Hz which further we will designate as f_0. For further calculations we will need to know the speed of rotation of the separator, which can be calculated according to the following expression:

$$f_c = \tfrac{1}{2} f_o (1 - \frac{d_{ш}}{d_c} \cos\alpha), \qquad (1)$$

Where $d_{ш}$ – is diameter of rotation bodies (balls), d_c – is diameter of separator, α is angle of bearing tension, in our case it is equal to zero. Based on our bearing SKF 6207-2z, f_c ranges from 12 to 14 Hz. However, in our case, it is not this chastota itself that is of interest, but the frequency of rolling of the bodies of revolution (bearing balls) through the outer ring of the separator, which is defined as:

$$f_H = f_C \cdot z, \qquad (2)$$

Where z - is the number of bodies of rotation, in our case 9, which gives a frequency of 117 Hz. It is in the vicinity of this frequency that we have to control the vibrations on the bearing, to estimate the test (diagnose) fault.

3. Results of the experiment and discussion

The conducted experiment can be estimated both from the signal itself, which will be presented below, also by the amplitude spectrum of the signal in the frequency range close to the value calculated by us. Figure 4 shows the vibration signal removed from the vibration inverter on the is-right bearing, as well as the part of the signal that can be obtained from it if you subtract the RMS signal. For the numerical evaluation of the instrument vibration signal RMS, we used the following integral indicator:

$$VA = \sqrt{\frac{1}{T_v}\left(\int_0^{T_v} \frac{da(t)}{dt}^2 dt\right)} \quad (3)$$

Where $a(t)$ - is the vibration signal itself (vibration acceleration), and T_v - the period of signal observation.

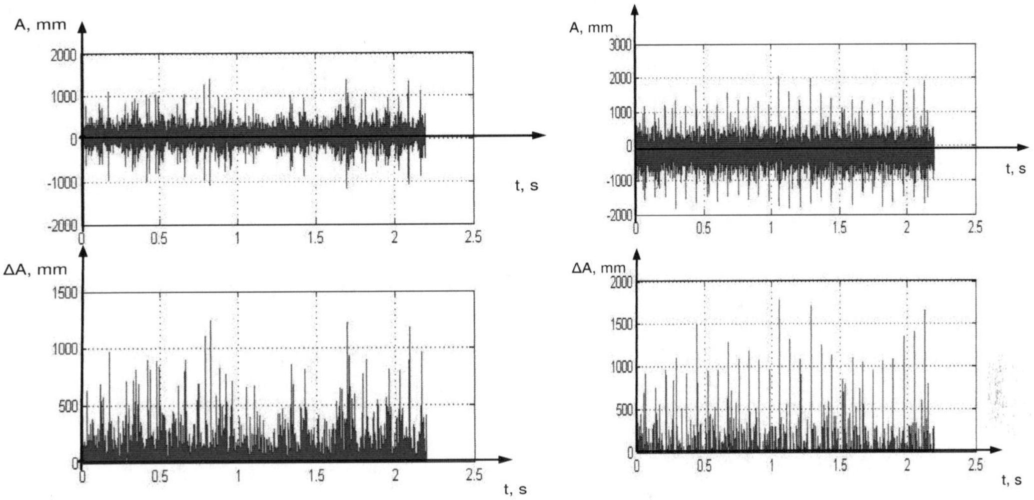

Figure 4. Vibration acceleration and its peak values: a) serviceable bearing, b) faulty bearing.

As can be seen from Figure 4 a), the visual analysis of what is happening on the test bearing is not amenable, however, it is of interest between the vibration signals of a serviceable and a unserviceable bearing, in our case the vibration signal from a unserviceable bearing is shown in Figure 4 b). A comparison of Figures 4 a) and 4 b) allows us to see only the fact that the vibrations on a unserviceable bearing are higher (greater) in amplitude, but how much more they are from the figure is not clear, just as it is not clear what exactly leads to this difference in signals . To determine the cause of the amplitude difference of the oscillation signals on a serviceable and a unserviceable bearing, it is convenient to consider the amplitude spectrum of the signal taken from a unserviceable bearing, which we presented in Fig. 5 for the case of an experiment without imbalance.

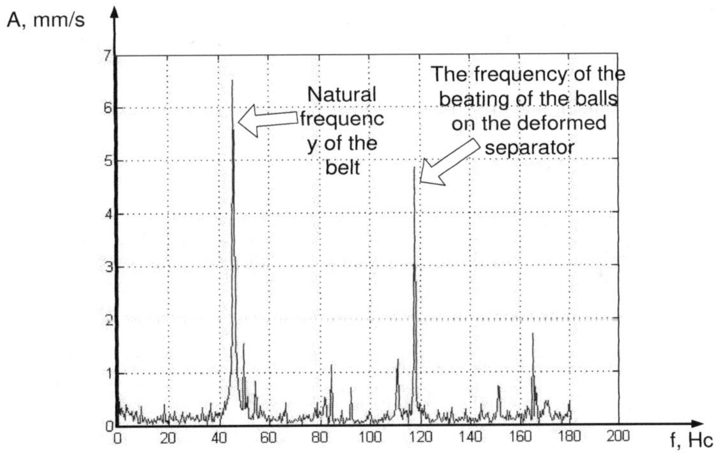

Figure 5. Spectrum of signal removed from unserviceable bearing, without unbalance

As can be seen from Figure 5, two main frequencies of amplitude peak are present in the frequency range of the signal. The frequency is 48 Hz, which is associated with the vibration of the drive belt and the frequency of 116 Hz, which, according to our calculations, is associated with a bearing defect. As can be seen from Figure 5, the amplitude spectrum power here is greatest. In order to assess the impact of the fault on the RMS vibration signal, we passed the measured signal through a frequency filter, with a window from 100 to 140 Hz, and calculated the value of the signal energy using the expression (3). Results for the serviceable VA=793 mm/s2 bearing, and for a case of the unserviceable bearing were VA=5823 mm/s2, gain of energy of vibrations much more seven times that is not visible from comparison of figures 4 a) and 4 b). For order to evaluate the operation of this approach in conditions close to processing conditions, we will consider the amplitude spectrum of the signal removed from the unit with unbalance on the rotor (see Figure 6).

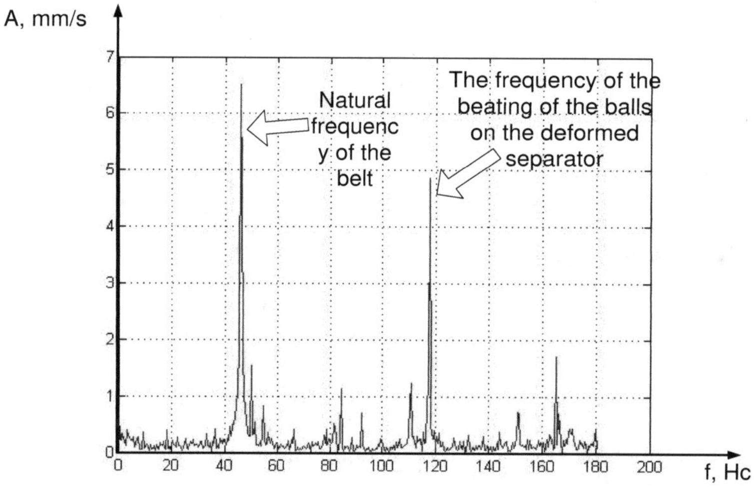

Figure 6. Spectrum of the signal recorded taking into account the introduced imbalance

As can be seen from Figure 6 as compared to Figure 5, there are significant run-outs at the rotor speed, but the run-outs associated with the diagnosed fault, if changed, are extremely insignificant.

4. Conclusions

The article presents a method for solving the problems of diagnosing bearings of spindle assemblies of metal-cutting machines that have defects in the outer rings of the separator. The paper proposes approaches to such an analysis based on the determination of frequency ranges and fixing faults and estimates of the RMS of the vibration signal in this frequency range. Studies have shown that the difference between the serviceable and unserviceable bearings in a predetermined private window of the oscillatory signal can be more than seven times, which, in our opinion, is quite enough for the correct determination of the occurrence of malfunctions.

This study was conducted on behalf of the Russian Foundation for Basic Research (RFBR) (Grant no. 19-08-00022).

References

[1] Lapshin V, Moiseev D, Minakov V 2019 *AIP Conference Proceedings*. Vol. 2188.
[2] Zakovorotny V L et al. 2019 *E3S Web of Conferences*. EDP Sciences. Vol. 104.
[3] Lapshin V P et al. 2020 *Russian Engineering Research*. Vol. 40. P. 259-265.

[4] Lapshin V P, Babenko T S, Moiseev D V 2018 *International Conference on Industrial Engineering*. Springer, Cham. P. 853-859.

[5] Kozochkin M P et al. 2018 IEEE International Conference *Quality Management, Transport and Information Security, Information Technologies (IT&QM&IS)*. P. 458-461.

[6] Sabirov F S, Vainer L G, Rivkin A V 2015 *Russian Engineering Research*. Vol. 35. P. 458-461.

[7] Porvatov A N et al. 2015 *Mechanics & Industry*. Vol. 16. P. 707.

Selective suppression of amplitudes of frictional oscillations

V I Kolesnikov[1,*], V V Shapovalov[2], A L Ozyabkin[2], M M Shestakov[2], R A Kornienko[2], T D Pertseva[3] and A M Petrik[3,4]

[1] Department of Theoretical mechanics, Rostov State Transport University, Rostov-on-Don, pl. Rostovskogo Strelkovogo Polka Narodnogo Opolcheniya, 2, 344038, Russia
[2] Department of Transport machines and Tribotechics, Rostov State Transport University, Rostov-on-Don, pl. Rostovskogo Strelkovogo Polka Narodnogo Opolcheniya, 2, 344038, Russia
[3] Faculty of road construction machines, Rostov State Transport University, Rostov-on-Don, pl. Rostovskogo Strelkovogo Polka Narodnogo Opolcheniya, 2, 344038, Russia
[4] Author to whom any correspondence should be addressed,

*e-mail: adad98999@mail.ru

Abstract. Maintenance of sanitary codes and standards during operation of railway sorting stations, in particular, standards which concern noise levels, is now the most pressing task. The analysis of noise has been carried out during operation of brake beams of car hump retarders with car wheelsets; the most problem frequency ranges have been revealed. On the basis of modelling bench tests with application of techniques of physical and mathematical modelling and tribospectral identification of friction processes we carried out selection of noise-suppressing cladding material which would reduce the level of acoustic noise at interaction of metal surfaces. Results of operational trials have confirmed efficiency of the chosen material in noise reduction.

1. Introduction

People have always lived in the world of sounds and noise. In nature loud sounds are rare; noise is rather weak and short. Normal noise of environment surrounding the person varies within 35-60 dB, where dB is a measuring unit of noise or sound pressure. The combination of sound stimuli gives time to a person for an estimation of their character and response. Growth of urban population, and construction of the industrial enterprises in the former suburbs of urban areas cause increase in levels of sound pressure [1] to acoustic organs of people , 1.5-2 times exceeding admissible norms.. Sounds and noise of high-power expose people's health, and can cause pain; they can also affect hearing aid and nerve centres, and provoke adverse effects on performance of cardiovascular system and other human organs. The above-listed enterprises include stations and railway junctions that are characterized by increased rolling noise and vibration levels caused by wheel and induced at the wheel-rail contact. There are some effective ways to reduce the harmful effects on individuals, for instance, constructive solutions for damping the track structure and operating brake marshalling complexes at mechanized marshalling yards.

These days the most technologically advanced countries are actively engaged in improving the eco-friendliness of technical systems, including reducing noise and vibration [2-10]. For example, "Living Well, Within the Limits of Our Planet" Environment Action Program [11], which is guiding European

environment policy until 2020, sets out commitments to reduce noise pollution in the European Union and to reach the World Health Organization-recommended level by 2020.
Due to the high level of noise and vibrations induced by ICEs, in particular, the engines of the MI 26 helicopter, the suppression of these dynamic processes is a complex, yet an extremely important and urgent task. Currently, the below-mentioned methodology has become the ground for creating a set of structural and technological measures aimed at improving the technical, economic and environmental characteristics of the MI 26 helicopter power actuator.

The research was carried out as part of Federal Targeted Program for Research and Development in Priority Areas of Development of the Russian Scientific and Technological Complex for 2014-2020 and financially supported by the Ministry of Science and Higher Education of the Russian Federation (Project Identifier RFMEFI60718X0203).

ELPA company has developed BREMEX-ANNSYS system [12] for noise level reduction up to 30 dB at sorting stations. The system works in such a way that special composite material DBM for dry braking is applied precisely to car wheels or to where braking takes place before there is a frictional contact between them. This eliminates oscillations in a range of resonant frequencies. Since a considerable quantity of energy is released in the form of heat during friction between wheels and rails, the auxiliary component of the composite material evaporates, whereas the working component (more often it is aluminium) creates an abrading brake film which removes a grinding sound in its source. BREMEX-ANNSYS system has some drawbacks. First, it has microprocessor control which considerably increases capital investments to the equipment of one brake position (300 thousand euro); then, it is impossible to use this system under conditions of considerable environmental change: air temperatures (from-50 °C to +50 °C), wind loading P (from P_0 to P_{max}), atmospheric precipitation in the form of rain and snow; next, evaporations of auxiliary component reduce ecological safety; finally, use of the consumable material (modifier) is exclusively made by ELPA company.

2. Carrying out of tests

The reason of the high noise level appearing during car hump retarder of mechanised sorting station is resonant oscillations and frictional self-oscillations occurring during interaction of micro-and macroroughnesses of side faces of wheels of freight cars with operating surfaces of brake beams. In this case it is possible to present micro-and macroroughnesses in the form of a model of individual roughness (figure 1), consisting of two dynamic equivalent single mass systems with natural frequencies $\omega_1 = \sqrt{\frac{C_1}{m_1}}$ and $\omega_2 = \sqrt{\frac{C_2}{m_2}}$. In case of coincidence of frequency of dynamic influence of system and one of the frequencies ω_1 or ω_2, there are resonant fluctuations of lumped masses m_1 or m_2 (or m_1 and m_2).

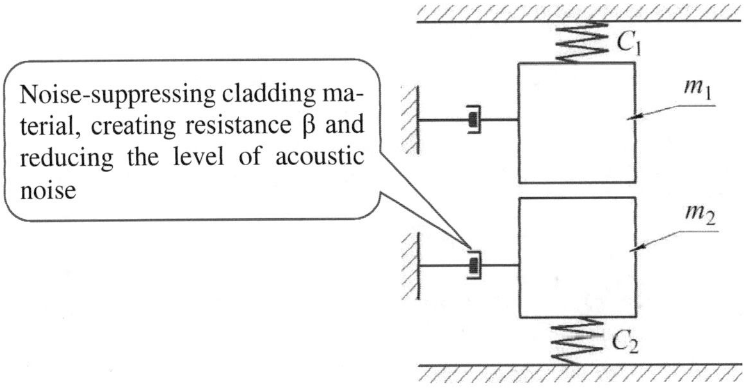

Figure 1. Model of individual roughness

The principal cause of origination of frictional self-oscillations consists in nonlinear dependence of frictional forces F_{ff} that happen due to more than 40 factors. To reveal the reasons of origination of oscillations and selection of noise-suppressing cladding material we carried out field investigations and modelling bench tests.

We used a special device EKOFIZIKA-110A (ecophysics – 110A) to check in levels of sound pressure during performance of a car hump retarder at the mechanised sorting station. To reveal frequency range where level of sound pressure exceeds sanitary codes and standards, we carried out tests using dictaphone Olympus WS-853 and software that performs octave spectral analysis.

As a result of 1/12-octave analysis (figure 2) of the noise appearing during interaction of wheels of cars with brake beams at mechanised hump yards of the railway stations Bataysk (the North Caucasian railway), it has been established that the high noise level is observed on the frequencies exceeding 2 000 Hz, and acoustic waves of high power are observed on center resonant frequencies of 2 986 and 5 625 Hz.

For carrying out of bench tests on selection of lubrication and elimination of resonant oscillations and frictional self-oscillations during interaction of brake tyres of a car hump retarder with wheels of cars, we used physical and mathematical modelling [13] and engineering test bed «rolling stock - track», executed in geometrical scale C_l = 5. Thus, we reproduced similar loading-speed conditions of interaction on the scale of forces of normal and tangential interactions $C_N = C_l^2$, sliding velocities $C_{Vc} = 1$, contact pressure $C_Q = 1$, frequencies of natural fluctuations $C_\omega = 1$ both on micro - and on macro - levels of frictional interaction.

Traditionally, friction factor is represented in the form of the static relation of average frictional force to average force of safe pressure $\mu = \frac{F_{ff}}{N}$; unlike that, the result of interaction of forces of friction engagement is estimated by complex value [13] on every frequency of harmonious fluctuation with frequency ω in the form of transfer function

$$W(i\omega) = \frac{S_\tau(i\omega)\cdot S_N(-i\omega)}{|S_N(i\omega)|^2} = \frac{S_{\tau N}(i\omega)}{S_{NN}(\omega)} = A(\omega)\cdot e^{i\varphi(\omega)} = P(\omega) + iQ(\omega), \quad (1)$$

where $S_{NN}(\omega)$ is autospectral function of normal influence in frictional contact taking into consideration the uncontrollable power noise disturbing stationary movements of the system;

$S_{\tau N}(i\omega)$ is the spectral function of mutual tangential and normal interaction taking into consideration uncontrollable power noise, properties of triboconditions and the external conditions influencing the system;

$A(\omega)$ is amplitude-frequency characteristic

$$A(\omega) = |W(i\omega)| = \sqrt{P^2(\omega) + Q^2(\omega)};$$

$\varphi(\omega)$ is phase and frequency characteristic

$$\varphi(\omega) = \operatorname{arctg}(Q(\omega)/P(\omega));$$

$P(\omega)$ is material frequency characteristic that describes elastic-inertial properties of the system, or potential kinetic energy

$$P(\omega) = Re\{W(i\omega)\} = A(\omega)\cdot\cos[\varphi(\omega)];$$

$Q(\omega)$ is imaginary frequency characteristic that describes dissipative losses in tribosystem, or function of energy dispersion

$$Q(\omega) = \operatorname{Im}\{W(i\omega)\} = A(\omega)\cdot\sin[\varphi(\omega)];$$

ω is periodic signal frequency.

The result of calculations of frequency transfer function (1) is the set of amplitude and phase-frequency characteristics (figure 3) that correspond to a variety of statuses of frictional contact.

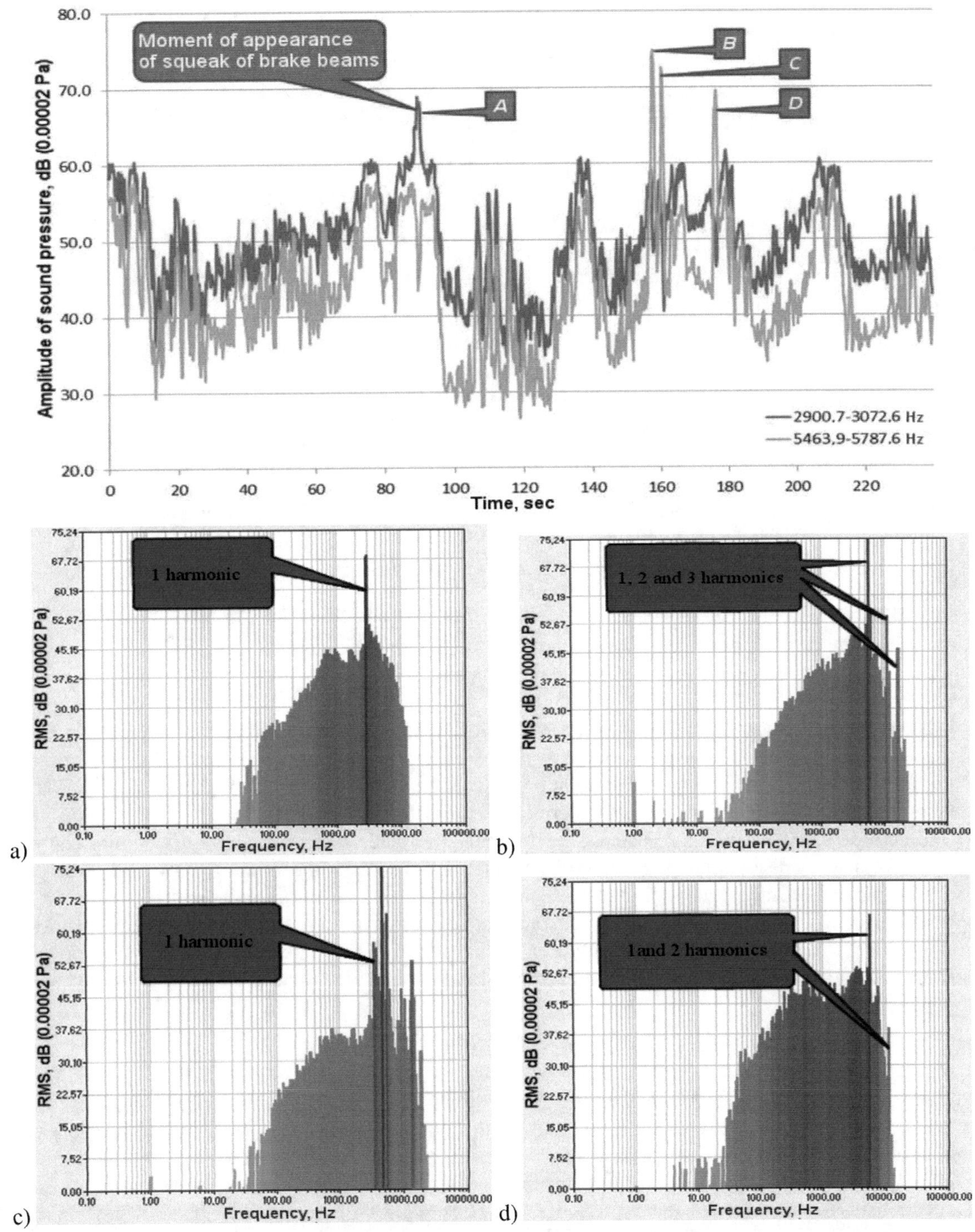

Figure 2. Basic 1/12-octave frequency ranges of vibroacoustic noise, appearing during interaction of wheelsets with brake tyres on the mechanised hump yards:
a – on frequency of 2 986 Hz; b-d – on frequency of 5 625 Hz

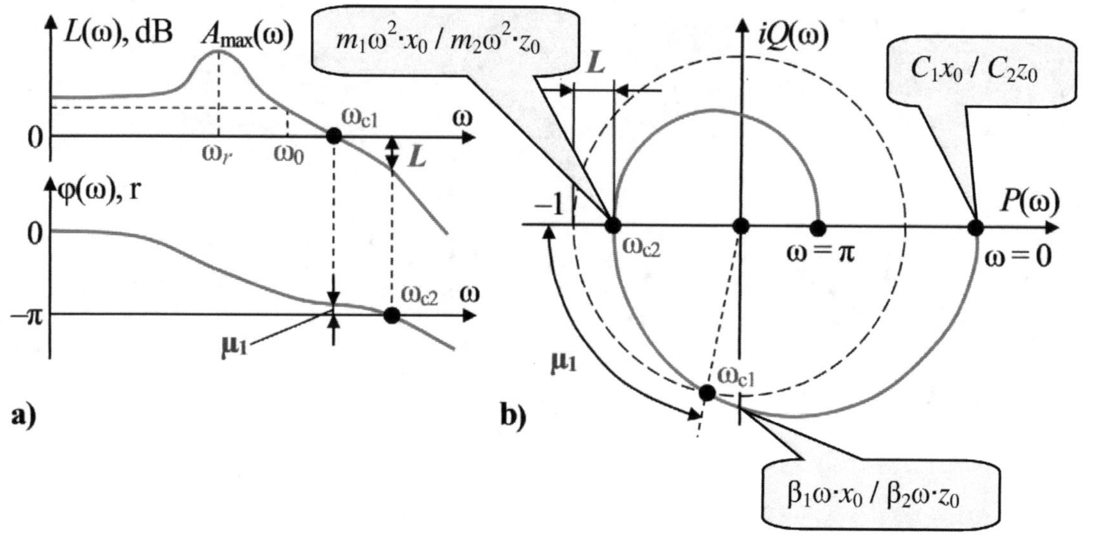

Figure 3. Frequency characteristics of tribosystem: a – amplitude, and phase; b - amplitude-phase, where integrated values: m_1, m_2 are masses of the active microvolumes participating in friction [kg]; C_1, C_2 are factors of elastic properties of contact $[\frac{N}{m}]$; β_1, β_2 are factors of resistance to fluctuations $[\frac{Nanosecond}{m}]$; ω is frequency of fluctuations [c^{-1}]; z_0 is initial displacement in a normal direction; x_0 is initial tangential displacement; L is factor of a stock of stability on amplitude; μ_1 is factor of a stock of stability on a phase

The dynamic friction factor W ($i\omega$) varies from 0 to ∞ depending on prevalence of elastic or dissipative component, presence or suppression of dynamic processes. For example, in case of realisation of modes of athermal or thermal adhesion, the friction factor tends to infinity, and in case of creation of resonant fluctuations it tents to zero (Tolstoi-Push effect).

Integrated estimation was suggested to identify friction and wear processes in frictional contact area and to select noise-suppressing cladding material on the basis of the analysis of amplitude and phase-frequency characteristics (figure 3). This integrated estimation allows estimating a ratio of elastic-inertial and dissipative forces of frictional interaction, stability, and irreversibility conditions in the contact area. For example, the integrated estimation of a dissipative component of friction I_Q, defining dissipative properties of a mechanical subsystem and friction process as dynamic connection [13]

$$I_Q = \int_{\omega_i}^{\omega_j} |Q(\omega)| d\omega,$$

where ω_i, ω_j are boundary frequencies during octave analysis or 1/3-, 1/12-, 1/24- octave analysis.

Using above-mentioned research techniques of the nonlinear processes taking place in frictional contact area including accompanying frictional self-oscillations, we carried out a model optimisation of five variants of noise-suppressing cladding materials. After the analysis of the received data we selected the composition based on a natural thermoplastic-adhesive. A dissipative component of frictional interaction of the given composition on problem CPB frequency ranges of the frictional self-oscillations appearing during interaction of operating devices car hump retarder with wheel side faces of freight car wheel is on 25-30 % more than that of the other investigated variants. The given composition has been chosen as a basic one.

We have developed three schemes of technical solution to creation of process equipment for application of thermoplastic-adhesive. As the optimum scheme of the device for cladding a wheel side face we chose the design with an arrangement of briquettes of modifiers in the body of operating devices of standard car hump retarders (figure 4).

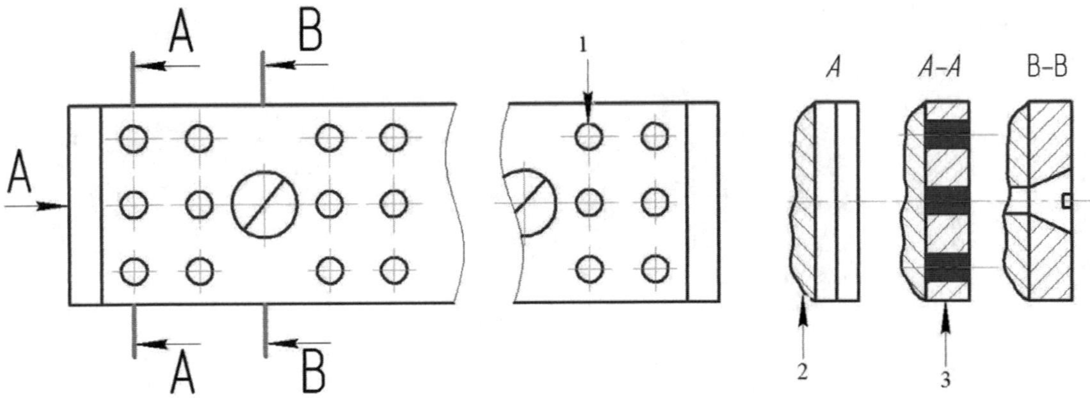

Figure 4. Brake shoe: 1 - noise-suppressing cladding material; 2 - beam case of retarder; 3 - friction facing

Modifying process is carried out by feeding noise-suppressing cladding material from the cavities which are located in bodies of a friction pair (for example, in a friction facing body) during wear-out of the body in contact process with another body (with its operating surface, for example, with a lateral surface of a wheel of a freight car). The necessary quantity of noise-suppressing cladding material is determined by intensity of wear process of friction surface of the first body (with cavities and with noise-suppressing cladding material). In its turn, the intensity of wear process is determined by a ratio of the sectional area of cavities with noise-suppressing cladding material S_1 to the contact area of material of the first body $S_2 = S_\Sigma - S_1$, where S_Σ is the area of a friction surface.

Real-life tests of the experimental device on noise reduction at the mechanised hump yards and on application of the modifier to a friction surface on lateral sides of wheelsets were carried out with use of empty and loaded freight cars with weight of 69-70 tons. As a result of interaction of wheelsets of freight cars with brake beams, we fixed formation of a modifying covering on brake beams of the device (figure 5) and egress of noise-suppressing cladding material (NSCM) from special accumulators on lateral sides of wheelsets (figure 6).

After the device applied successfully noise-suppressing cladding material on wheelsets of the freight cars, they retarded in a regular mode from a hump yard. By means of EKOFIZIKA-110A we made the octave analysis of the noise appearing within a distance of 2 and 7.5 m from retarder positions (table 1) during sorting of freight cars with modifiers of friction surface applied to lateral sides of wheelsets.

Table 1. The octave analysis of the noise appearing during operation of car hump retarder

	Frequency, Hz (average of 10 tests)									
	31.5	**63**	**125**	**250**	**500**	**1,000**	**2,000**	**4,000**	**8,000**	**16,000**
L = 2 m	56.1	53.4	49.9	48.7	52.3	49.9	51.2	49.4	46.8	40.8
L = 7.5 m	41.9	44.3	42.8	47.7	52.7	44.6	40.4	41.9	44.3	42.8

IOP Conf. Series: Materials Science and Engineering **900** (2020) 012005 doi:10.1088/1757-899X/900/1/012005

Figure 5. Configuration of system of axis drive of noise-suppressing cladding material (NSCM) on tracks of Bataysk permanent way division

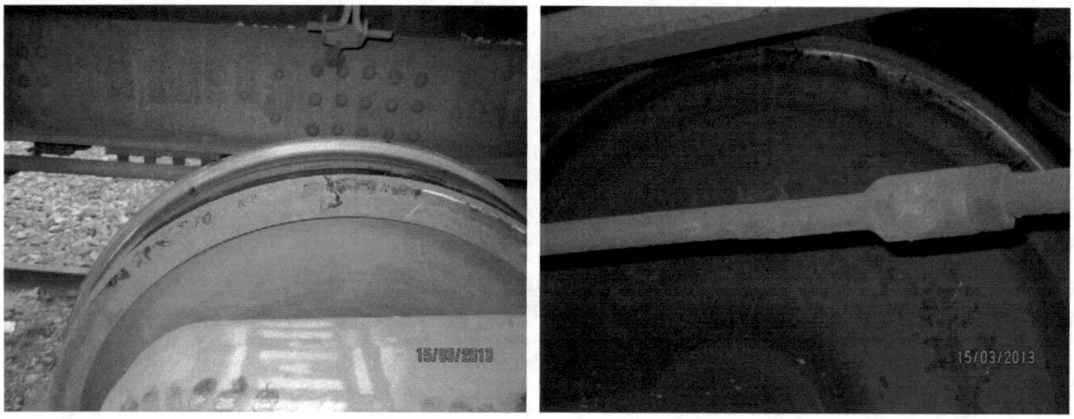

Figure 6. Modifier of a friction surface on lateral sides of the freight car wheels

It is established that at the admission of the test car to all three retarder positions of the mechanised hump yard the noise suppression has been revealed on all frequency ranges up to the level which is not exceeding sanitary codes and standards.

3. Conclusions

The following conclusions have been drawn as a result of the analysis of the operational trials.

1) The experimental design of the device provides application of necessary quantity of noise-suppressing cladding material onto a side face of car wheels; it provides reduction of noise (which appears at interaction of brake beams with disks of wheelsets) to the values which are not exceeding admissible sanitary codes and standards.

2) Due to an intake device and absence of connections in a horizontal plane, the complex design well adjusts (is guided) to the freight car wheels which side face is modified by the device for reduction of noise level.

3) Due to application of the experimental device of an axis drive system of noise-suppressing cladding material before the first brake position we can evidence a drop of levels of sound pressure and noise from 85 … 97 dB to 40 … 52 dB in octave frequency bands to the values regulated by sanitary codes and standards.

IOP Conf. Series: Materials Science and Engineering **900** (2020) 012005 doi:10.1088/1757-899X/900/1/012005

4) The rotaprint and contact method of application of noise-suppressing cladding material provides reliable and effective operation of the device in a wide temperature range (from-50 °C to +50 °C) and presence of wind loading, providing thus ecological cleanliness as a result of absence of effects of greasing of surfaces of rails and rolling stock.

References

[1] Salzmann H 1998 *Eisenbahningenieur* **4** pp 60–64

[2] https://www.getzner.com Getzner's website (Germany)

[3] Thompson D and Jones C 2002 *Railway Gazette Int.* **7** pp 363–366

[4] Dumitriu M and Cruceanu I C 2017 *J. of Engineering Science and Technology Review* **10(6)** pp 87-95.

[5] Altenbaher B, Goltnik D and Rosi B 2015 T*ransport problems* **10(2)** pp 5-14

[6] NVW editorial staff 2018 *Noise and vibration Worldwide* **49(4)** pp 136-139

[7] Zvolenský P, Grencik J, Pultznerová, A and Kašiar Ľ 2017 *MATEC Web of Conferences* vol 107

[8] Zhao C, Wang P, Yi Q, Sheng X and Lu J 2018 *J. of Low Frequency Noise, Vibration and Active Control* **37(4)** pp 896-910

[9] Zhao C and Wang P 2015 *Shock and Vibration* **2015** pp 1-12

[10] Marjani S R and Younesian D 2019 *Shock and Vibration* **2019** pp 1-13

[11] Potočnik J 2013 *European Commission. General Union Environment Action Programme to 2020* (Brussels)

[12] https://www.virtualmarket.innotrans.com/en/BREMEX-ANNSYS,p1024248 BREMEX-ANNSYS: ELPA Virtual Market Place, Exhibitors & Products (date of the application: 19.03.2017)

[13] Shapovalov V V, Shestakov M M et al 2018 Patent No. 2674391 of the Russian Federation, IPC V61K 7/02. *The Method to Suppress Noise at the Interaction between the Working Bodies of Car Retarding Mechanism and the Side Surfaces of the Freight Car Wheels*

IOP Conf. Series: Materials Science and Engineering **900** (2020) 012006 doi:10.1088/1757-899X/900/1/012006

Mobile robot for running exercises

A I Izyumov, E V Vasilyeva*, S V Lobanov, A E Sergienko and E A Alentsov

Don State Technical University, Rostov-on-don, Russia

*e-mail: catrinfox@yandex.ru

Abstract. Progress does not stand still, and machines are involved in the process of training professional athletes, where the greatest restriction is economic in nature - simulators are not available to a wide range of users. The field of running sports is currently practically not provided with robotic support in such a way that with the help of machines it was possible to implement the process of training an athlete. This article is devoted to the construction of the concept and principles of the implementation of a mobile robot for performing running exercises.

1. Introduction

Due to the intensive growth of technological progress, modern robotics as a science, in one way or another, affects most areas of society socialization. The expansion of areas of application of robotics occurs constantly, simultaneously with the growth of human needs, and today one of the most relevant branches of development is the field of sports robotics.

With the help of robots, sports equipment is tested, control and accounting processes are performed to record the results of sports competitions, and finally, competitive training is carried out, aimed at performing basic exercises that can eventually improve the performance of professional athletes.

This article does not consider the simplest mechanical devices, such as ball-serving machines in tennis and baseball, which serve with different forces and at different angles in a given range. A significant disadvantage of such robots is their one-sidedness – such a device will not respond to a counter feed. Therefore, of the already implemented solutions, manipulator robotics is usually used [1, 2] (figure 1 a, b).

a) б)

Figure 1. Training robots-manipulators for: a) table tennis; b) volleyball

IOP Conf. Series: Materials Science and Engineering **900** (2020) 012006 doi:10.1088/1757-899X/900/1/012006

2. Concept

It is proposed to develop a mobile robot for novice athletes, competing with a group or individual athletic running segments with a set time when moving around the stadium. In the course of calculations of technological parameters of components of the prototype being developed, it was decided to consider the task of using a mobile runner robot as appropriate [3].

Table 1. Material development costs

Name of materials, Ed. ed.	Units	Quantity	Price per unit, RUB.	The total amount of
Arduino UNO platform	Pcs	1	299	299
Photodiodes TEMD5080X01	Pcs.	4	37	148
Motor driver L293D	Pcs	1	220	220
El. Engine MY6812	Pcs.	2	2200	4400
Battery RED ENERGY DS 1205	Pcs.	1	2125	2125
ResistorS2-23 2W-0R22-5%	Pcs.	4	8	32
Housing (aluminum 3x1200x3000 mm 1105anr)	Pcs	2	210	420
Housing treatment	Pcs	1	500	500
Wheel 65 mm	Pcs	4	160	640
Total:				8784

The appearance of the mobile robot runner is a frame, lower and upper parts of the body. The frame bearing the entire load from the mass of the engine, battery, controller and sensors is shown in figure 2, the location of the components is shown in figure 3.

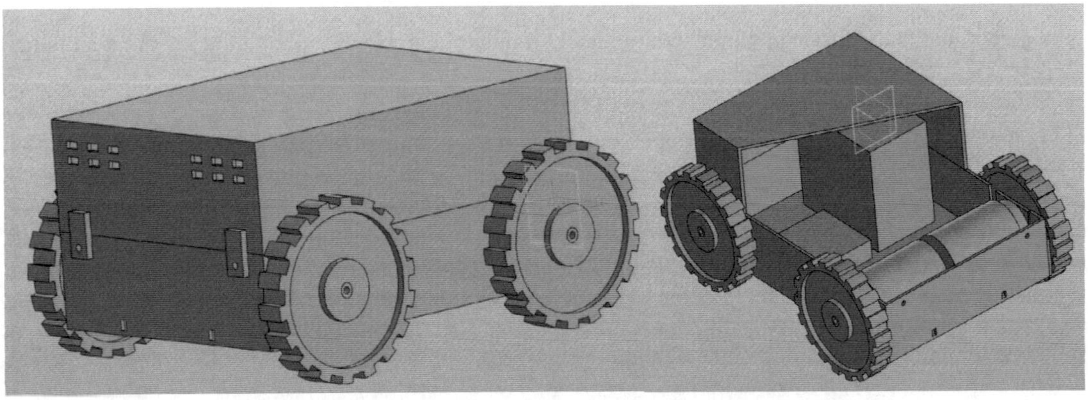

Figure 2. Appearance of the prototype under developed

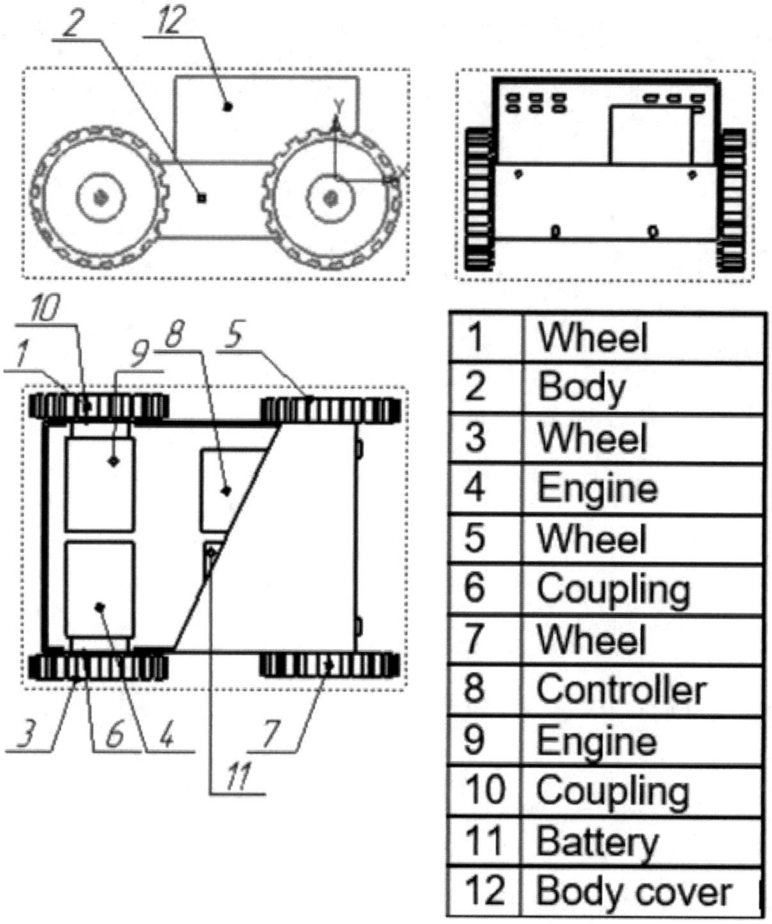

Figure 3. Location of components of the prototype under development

The dimensions of the mobile robot are not stated in this article, because they may change during the actual process of device assembly.

The dimensions of the housing do not exceed the parameters of the DW: 276 × 197 × 126 (mm).

The fundamental principle of sports competitions is to compete with the opponent. This kind of work encourages you to improve your own performance, and the prototype you are developing acts as an opponent on the treadmill. The mobile app sets the distance and time to cover it. The speed of movement of the device on the treadmill depends on these parameters. The athlete, accordingly, follows the robot for the purpose of overtaking, or accompanies the robot in order to keep within the range of the set time. The robot's trajectory is determined by fixing the sensor dividing strip between the stadium's running tracks. At the end of the workout, the mobile app saves all parameters (distance, time, speed, and the athlete's heart rate (if you have a fitness bracelet connected to the app via Bluetooth)) it also displays summary statistics in comparison with the current standards of athletes' training [4] .

The estimated power of the prototype being developed, taking into account the implementation of the process of training non-professional athletes, should allow to develop a speed sufficient to meet the standard "candidate for master of sports" inclusively at a distance of up to 600 m, as well as passing long distances of up to 5 km by athletes, inclusive [5].

Discharge standards for women and men are presented in tables 2 and 3, respectively.

Table 2. Digit standards for men at distances up to 600 m
(Minutes: seconds, fractions of seconds)

Distance	Circle	Kms Circle	I	II	III	I (u)	II (y)	III (y)
30m	-	-	-	-	4,2	4,4	4,7	5,0
50m	-	-	6,1	6,3	6,6	7,0	7,4	8,0
60m	-	6,8	7,1	7,4	7,8	8,2	8,7	9,3
100m	-	10,7	11,2	11,8	12,6	13,5	14,4	15,4
200m	-	22,0	23,0	24,3	26,0	28,0	30,5	34,0
200m	200m	22,4	23,4	24,7	26,4	28,4	30,9	34,4
300m	-	34,5	36,2	38,7	41,6	45,2	49,0	53,0
400m	400m	49,5	52,0	56,0	1:00,0	1:05,0	1:10,5	1:16,2
400m	200m	50,3	52,8	56,4	1:00,8	1:05,8	1:11,3	1:17,0
600m	400m	1:22,5	1:27,5	1:33,5	1:40,5	1:47,5	1:54,5	2:05,0
600m	200m	1:23,7	1:28,7	1:34,7	1:41,7	1:48,7	1:55,7	2:06,2

Table 3. Level standards for women at distances up to 600 m
(Minutes: seconds, fractions of seconds)

Distance	Circle	kms Circle	I	II	III	I (u)	II (y)	III (y)
30m	-	-	-	-	4,9	5,1	5,3	5,6
50m	-	-	6,9	7,3	7,7	8,2	8,6	9,3
60m	-	7,6	8,0	8,4	8,9	9,4	9,9	10,5
100m	-	12,4	13,2	14,1	15,1	16,1	17,2	18,4
200m	-	25,5	27,1	29,2	31,5	33,9	36,3	39,7
200m	200m	25,9	27,5	29,6	31,9	34,2	36,9	39,9
300m	-	40,5	42,8	45,5	49,5	53,6	58,0	1:03,0
400m	400m	57.5	1:01,5	1:06,0	1:11,0	1:17,0	1:24,0	1:30,0
400m	200m	58,3	1:02,3	1:06,8	1:11,8	1:17,8	1:24,8	1:31,3
600m	400m	1:36,5	1:43,0	1:50,5	1:59,0	2:09,0	2:20,0	2:32,0
600m	200m	1:37.7	1:44,2	1:51,7	2:00,2	2:10,2	2:21,2	2:33,2

In the case of group training sessions, the robot calculates the average speed and starts moving in front of a group of athletes at the command of the mobile app. After the segment is completed, the robot stops moving.

3. Principle of operation

The robot moves using two motors. Turns are performed by increasing the speed on one of the engines. With the help of infrared sensors, the robot tracks, the position of the dividing line, which must remain between the sensors during all movement.

The structural diagram of the robot under development is shown in figure 4.

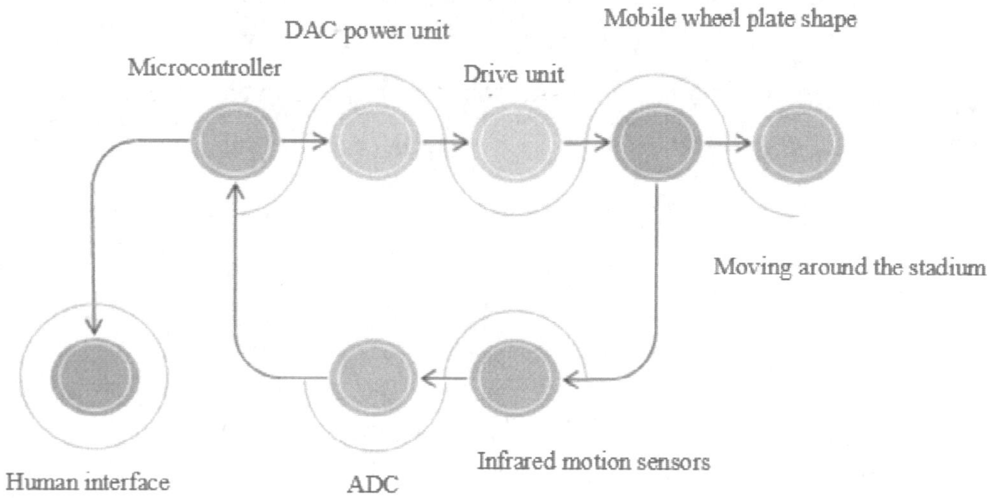

Figure 4. Block diagram

The robot is removed from its storage location using an application that provides a remote control (figure 5).

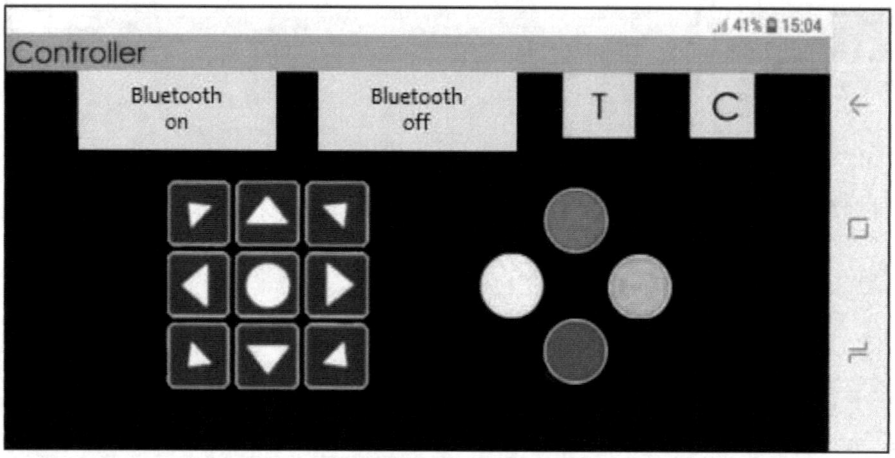

Figure 5. Navigation console for manually robot controlling

When the robot is put on the track and installed according to the location of the dividing lines between the sensors, the training process starts the app sets the length of the distance and the time it takes to complete it.

Conclusion

The advantages of design development are determined by the low cost relative to existing analogues and the versatility of its components. The device has a built-in function of automatic adjustment of the travel speed. The simplicity of the device reduces the risk of disabling the device and significantly increases the possibility of replacing individual elements in the event of a malfunction. Using the mobile app to control the robot makes it much easier to interact with the robot and allows you to concentrate on running. The project can be widely applied in both individual and group training sessions; the principle of competition allows you to achieve high results.

Of all the considered analogues, the best technical indicators – ease of design and programming, ease of user management, characterize the presented concept. The advantage is also economic efficiency, which is an important indicator for the development and further mass production. The project will bring innovations to sports robotics and opens up prospects for further developments in this field.

References

[1] https://futurist.ru/news/3207 (Electronic resource, last checked on 07.06.2020)
[2] https://obzor.press/hi-tech/2016032418466 (Electronic resource, last checked on 07.06.2020
[3] https://drive.google.com/drive/folders/1xsNctjaYu6UTubq1BSi5-fxX21KtAKEg (Electronic resource, last checked on 07.06.2020)
[4] https://vitablog.ru/dvizhenie/professionalnyy-beg.html (Electronic resource, last checked on 07.06.2020)
[5] https://marathonec.ru/razryady-normativy-po-begu/ (Electronic resource, last checked on 07.06.2020)

IOP Conf. Series: Materials Science and Engineering **900** (2020) 012007 doi:10.1088/1757-899X/900/1/012007

Correlation of the dynamic properties of the manufacturing process on metal-cutting machines and energy efficiency of the process

V L Zakovorotny, V E Gvindjiliya*

Don State Technical University, Rostov-on-Don, Russian Federation

*e-mail: sinedden@yandex.ru

Abstract. Increase of the efficiency of machine processing is related with minimizing the reduced costs for manufacturing a batch of parts while ensuring their quality. The efficiency of the cutting process is determined by minimizing the intensity of tool wear while ensuring the required quality metrics of parts. Optimization consists in determining the cutting speed in which the path traversed by the tool to critical wear and (or) loss of quality of parts will be maximized at manufacturing with fixed regimes. As a rule, the minimization of conditions of the wear rate correspond to the best quality metrics of parts. In addition, this condition does not contradict the performance requirement because when the cutting speed increases, there is always a tendency to reduce forces. The ways to reduce the intensity of wear are knew, but the influence of the dynamic state of the system on wear and loss of quality is not considered. Therefore, the determination of the correlation the dynamic properties of the cutting process and its energy efficiency, as well as the determination of the modes set by the trajectories of the machine's Executive elements, characterize the next stage of improving the cutting efficiency according to conditions of minimization of the work of forces on a given path. The results of research on this problem are presented in the article.

1. Introduction

The intensity of tool wear, as an increment of wear along the way, is one of the main parameters that characterize the manufacturing process [1-6]. This is due to the fact that the all the main characteristics and the characteristics of the quality of parts change [4 - 8] with growing wear. Therefore, the study of the influence of processing conditions, the lubricating and cooling medium, technological modes, tool geometry, vibrations, etc. on wear has been the subject of numerous studies [1 - 6]. The main attention focus on the thermal physics of processing, including non-equilibrium nonlinear thermodynamics of irreversible processes [1,2,5, 9-13] to clarify the influence of various factors on the wear intensity. In this regard, much attention focus on the influence of cutting speed on wear, since the speed with forces significantly affect the power of irreversible energy transformations in the cutting zone, which is the primary source of all physical phenomena (heat production, adhesion, diffusion processes, etc.) that affect wear. In particular, it is experimentally proved that when the speed increases, there is an optimal value at which the wear intensity is minimal. In this case, there is a transition from the prevailing adhesive wear to diffusion wear. Moreover, these conditions correspond to the best product quality indicators [1,2,4,5]. It is also shown that the wear depends on the power emitted in the area of convergence of the tool faces with the workpiece and chips [8 - 14]. On this basis, mathematical models are proposed for describing tool wear using Volterra integral equations with respect to the phase trajectory of the power of irreversible energy transformations in work [15, 16]. In turn, the power emitted in the cutting zone depends on the dynamic properties of processing and it changes during the evolution of the cutting system due to changes of the parameters of the dynamic link formed by cutting [15, 16]. The article develops the above provisions, deals with the problem of the elastic system and the choosing the parameters available for changing the dynamic link formed by cutting in which the work of the cutting process is minimized.

2. Mathematical modeling

Manufacturing with unchanged cutting conditions is usually considered when the influence of technological modes on the wear of cutting tools is studied [1-6]. In this case, for example, the influence on the wear of technological modes is analyzed. Therefore, we will focus on the mathematical formulation of the research problem for longitudinal turning of a non-deformable part of constant diameter and allowance. Deformations are considered from the tool side in a movable coordinate system, the movement of which is determined by the trajectories of the Executive elements, set as speeds $\mathbf{V} = \{V_1(l), V_2(l), V_3(l)\}^T \in \Re_{\mathbf{V}}^{(3)}$ are represented in the displacement function of saddle along the axis of rotation of the workpiece l. Since the turning of a shaft of constant diameter is considered in constant modes, then, $V_1(l) = 0$, $V_2(l) = const$, $V_3(l) = \pi D\Omega = const$ (fig. 1). We assume that $\mathbf{V} \subset \Im_{\mathbf{V}}$ where $\Im_{\mathbf{V}}$ is the set of permissible variations in the trajectories of the machine's actuating elements, and take into account the elastic deformations of the tool vertex relative to the carrier system in space $\mathbf{X} = \{X_1(l), X_2(l), X_3(l)\}^T \in \Re^{(3)}$. Then, given $\mathbf{X}$ and $\mathbf{V}$, technological modes (V_P, S_P, t_P) depend on the trajectories of the Executive elements $\mathbf{J} = \{V_1(t), V_2(t), V_3(t)\}^T \in \Re_{\mathbf{J}}^{(3)}$ and elastic deformations $\mathbf{X}$.

$$\begin{cases} t_P(t) = t_P^{(0)} - X_1; \\ S_P(t) = \int\limits_{t-T}^{t} \{V_2(\xi) - v_2(\xi)\} d\xi; \\ V_P(t) = \pi D\Omega - dX_3/dt, \end{cases} \tag{1}$$

where $t_P(t)$, $S_P(t)$, $V_P(t)$, is the depth, feed and cutting speed as functions of time; $T = (\Omega)^{-1}$ - the rotation time of the part, the inverse of the cyclic frequency Ω. Finally, based on previous studies [16-19], we present the elastic dynamic subsystem of the tool as spatial linear deformation displacements. In this case, it has potential properties and can be represented as

$$\mathbf{m}\frac{\mathbf{d^2X}}{\mathbf{dt^2}} + \mathbf{h}\frac{\mathbf{dX}}{\mathbf{dt}} + \mathbf{cX} = \mathbf{F}_\Sigma, \tag{2}$$

where $\boldsymbol{m} = [m_{s,k}]$, $m_{s,k} = m$, $npu{:}\, s = k$, $m_{s,k} = 0$, $npu{:}\, s \neq k$, $s,k = 1,2,3$ в kgf^2/mm, $h = \lfloor h_{s,k} \rfloor$ в kgf/mm, $c = \lfloor c_{s,k} \rfloor s,k = 1,2,3$ kg/mm are symmetric and positive-definite matrices of the inertial, velocity, and elastic coefficients. Forces F_Σ are determined $F_\Sigma = F + \Phi$ (fig.1). We will accept The hypothesis of F dependence on the cross section area S is accepted, the delay in the change of forces in relation to variations S take into account, their dependence on V_3, and the possibility of their representation as $F(t) = F_0(t)\{\chi_1, \chi_2, \chi_3\}^T$. Then

$$T_0 dF_0 / dt + F_0 = \rho\{1 + \mu \exp[-\alpha(V_3 - dX_3 / dt)]\}[t_P^{(0)} - X_1] \int_{t-T}^{t} \{V_2 - dX_2 / dt\} dt \ , \quad (3)$$

where T_0 is constant time delays of the forces [c]; ρ is the pressure at the front surfaces for low speeds kg/mm^2; μ is nondimensional coefficient; α is determines the decrease in the forces with increase in the speed. Constant T_0 can be represented in condition coordinates and regimes such as

$$T_0(S_P, t_p, V_3) = \frac{\kappa \zeta_c S_p(t) t_p(t)}{V_3(t)} \ , \quad (4)$$

where κ is the dimension of inverse of the tool vertex path required to establish a new stationary state [mm^{-1}]; ζ_c is the chip shrinkage coefficient (its broadening and thickening), depending on the properties of the material. The article takes into account that for Φ forces increase disproportionately as the back face of the tool approaches the workpiece. Then

$$\begin{cases} \Phi_1 = \rho_0\{ \int_{t-T}^{t} [V_2 - dX_2/dt]dt\}\exp[\alpha_1(-dX_1/dt)]; \\ \Phi_2 = \rho_0[t_P^{(0)} - X_1(t)]\exp\alpha_2[V_2 - dX_2/dt]; \quad \Phi_3 = k_T[\Phi_1 + \Phi_2], \end{cases} \quad (5)$$

where α_1, α_2 is steepness coefficient ρ 0 is parameters of rigidity; k_T is coefficient of mating friction between tool side and part. Equations (2-5) allow us to study the dynamic properties of the cutting process, including their changes from the technological modes set using dependencies (1) by the trajectories of the Executive elements. To solve this problem, it is necessary to build a model of evolutionary changes that depend on the trajectory of the power of irreversible transformations in the cutting zone. Previously, it was shown that changes in the system properties are manifested in the development of tool wear, as well as related changes in dynamic coupling parameters [14-16]. Therefore, when modeling, first, we will consider the evolution of wear, and second, we will take into account changes in the parameters of dynamic coupling

$$p_i = p_{i,0} + \Delta p_i(w^{(h)}) \quad (6)$$

where $p_{i,0}$ is the initial value of the parameter; $\Delta p_i(w^{(h)})$ is increment with increasing wear. The Case of limited wear of back face is considered. Wear is stimulated by forces. Take into account

$$w^{(h)}(t) = \beta_1\{A(t) + \beta_2 \int_0^t W(A - \xi) N(\xi) d\xi\} \quad (7)$$

where $w^{(h)}$ is wear in [mm]; β_1 is coefficient in [$(kg)^{-1}$], β_2 is dimensionless coefficient; $W(t - \varsigma)$ is the dimensionless kernel of the integral operator. When evaluating evolution, all coordinates in relative units $(t_P^{(0)} - X_1^*)$ are considered. This is power $N^{(t)}$, work $A^{(t)}$ and forces $\Phi_i^{(t)}$ $i = 1,2,3$. The core of the integral operator (7) takes into account two competing processes hidden in many interrelated physical phenomena that cause adaptation and degradation of the properties of the bodies in contact. It is convenient to present it as

$$W(t - \xi) = \{\exp[-(T_1)^{-1}(t - \xi)] + \mu_v \exp[-(T_2)^{-1}(\xi - t)]\} \quad (8)$$

where T_1, T_2 is the time settings in [s]; μ_v is the dimensionless parameter. The parameters T_1, T_2, μ_v, β_1, β_2 in (8) are determined based on experimental dynamics methods. Moreover, the power $N(t)$ is represented in a discrete form

$$\mathbf{A} = \{A[\Delta t], A[2\Delta t],... A[i\Delta t],... A[n\Delta t]\}; \mathbf{N} = \{N_1, N_2,...N_i,...N_n\}, N_i = \frac{1}{\Delta t}\int_{t_{i-1}}^{t_i} N(\xi)d\xi, \quad (9)$$

Irreversible transformations are analyzed only in the area where the back faces of the tool are paired with the workpiece at discrete time intervals. As such, it is convenient to consider the time of one revolution, that is

$$N_i = \frac{1}{T_i}\int_{t-T_i}^{t} \{\Phi_1(t)\frac{dX_1}{dt} + \Phi_2(t)[V_2 - \frac{dX_2}{dt}] + k_T[\Phi_1(t) + \Phi_2(t)][V_3 - \frac{dX_3}{dt}]\}dt \quad (10)$$

Systems (2) - (9) represent a nonlinear integro-differential system of equations, which is analyzed numerically.

Now the problem of minimizing the energy consumption for manufacturing a batch of parts while ensuring the required quality indicators is formulated. The sum of the processed surfaces of a batch of parts is presented as a single cutting path L (fig. 1).

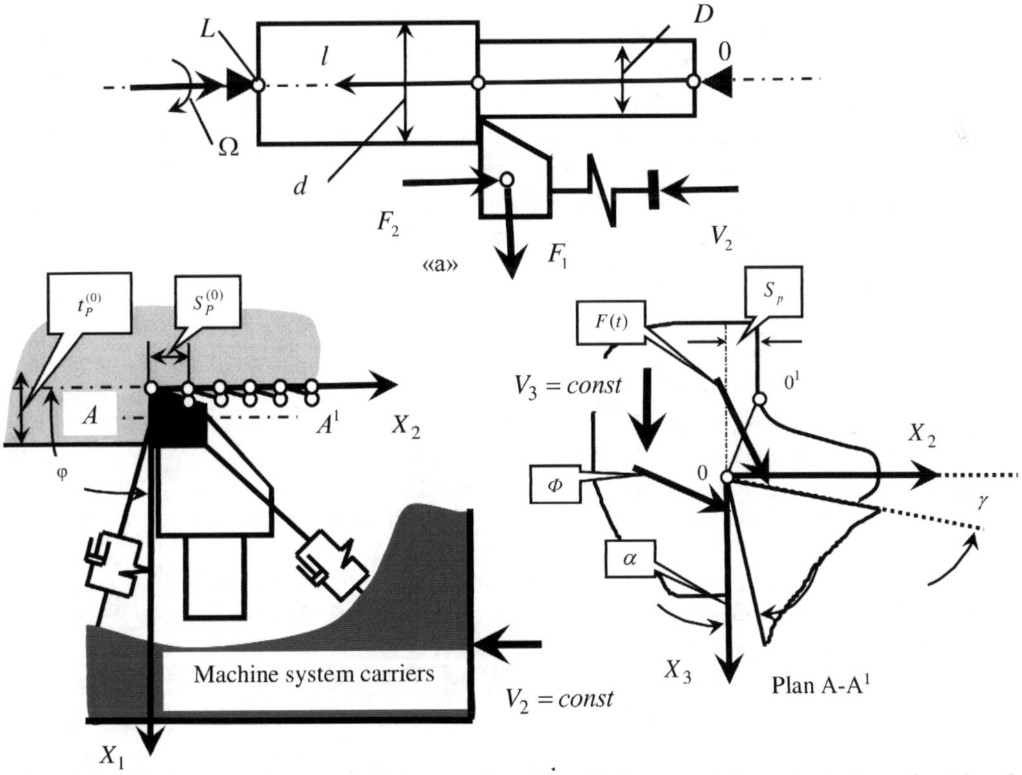

Figure 1. Dynamic system scheme: a – scheme of tool tip displacement; b – orientation of cutting forces and deformation displacement

Since longitudinal turning is considered, it is necessary to analyze two trajectories specified in the function of movement along the axis of rotation of the workpiece l. These are trajectories $V_2(l)$ and $V_3(l)$, which can be represented as discrete time vectors

$$\begin{cases} \mathbf{V_2} = \{V_2(\Delta t_1), V_2(\Delta t_2),..V_2(\Delta t_n)\}^T; \\ \mathbf{V_3} = \{V_3(\Delta t_1), V_3(\Delta t_2),..V_3(\Delta t_n)\}^T, \end{cases} \quad (11)$$

where $V_3(\Delta t_i) = \pi D \Omega(\Delta t_i)$. For $\mathbf{V_3}$ and D sequences of feed values $\mathbf{S_P} = \{S_P(\Delta t_1), S_P(\Delta t_2),..S_P(\Delta t_n)\}^T$ and $\mathbf{\Omega} = \{\Omega(\Delta t_1), \Omega(\Delta t_2),..\Omega(\Delta t_n)\}$ defined. Vectors (10) must be consistent with each other based on the requirements for surface roughness. Consistent trajectories characterize the set of acceptable variations, i.e.

$$\mathbf{V} = \{\mathbf{V_1}, \mathbf{V_2}, \mathbf{V_3}\}^{\mathbf{T}} \subset \Re_{\mathbf{V}} \Rightarrow \mathbf{V_2}/\mathbf{V_3} \le \boldsymbol{\delta_0} . \quad (12)$$

It is necessary to select such trajectories (10) for which the following conditions are met

$$\begin{cases} \sum_{i=1}^{i=n} S_P(\Delta t_i) = L; .... "a" \\ X_1 = const; .... "b" \\ \mathbf{A} = \{N_1 T_1, N_2 T_2,..N_n T_n\}, \quad \sum_{i=1}^{i=n} N_i T_i = \min ..."c" \end{cases} \quad (13)$$

where $N_i = \frac{1}{T}\int_{t_i - T}^{t_i} \{\Phi_1(t)\frac{dX_1}{dt} + \Phi_2(t)[V_{2,0} - \frac{dX_2}{dt}] + k_T[\Phi_1(t) + \Phi_2(t)][V_{3,0} - \frac{dX_3}{dt}]\}dt$, $\Omega(\Delta t_i) = (T_i)^{-1}$.

B (12) a – condition for processing a given group of surfaces (a group of parts); b – condition for the constancy of the diameter when the system parameters change evolutionarily; c – condition for minimizing work on a given path *l*. When solving the problem, we will use the most important properties of the evolution of the dynamic cutting system obtained earlier [14-16].

1) At the processing with constant technological modes, a complex nonlinear self-organization of the dynamic system is observed, which is accompanied by the formation of various attractive sets of forces and deformation displacements, including their bifurcations. The factor of "regulating" the formation of various attracting sets is the power of irreversible energy transformations in the areas of interface of the tool and workpiece faces. Therefore, with constant modes in the cutting process, their small variations and changes in the parameters of the dynamic cutting system can cause a significant restructuring of the evolutionary transformations, therefore, the total work on a given caliper movement. When the cutting speed is increased, the stability of the cutting process is first observed, and then conditions of parametric self-excitation are formed, which contribute to the loss of stability. This leads to the fact that when the speed is varied, there is a minimum of work on the specified caliper path, hence the minimum value of the wear intensity. Note that the constant path of the caliper corresponds to a different path and trajectory of the tool vertex, taking into account its deformation displacements.

2) All changes related to the work of forces that characterize irreversible transformations of the energy supplied to the cutting are also irreversible. For example, the development of wear and tear is irreversible. However, the properties of the dynamic coupling formed by cutting depend not only on wear, but also on the technological modes. Therefore, it is possible to coordinate the evolutionarily changing properties of the cutting process with changes in technological parameters set by the trajectories of the machine's Executive elements.

3) The intensity of wear depends on the power of irreversible transformations, which stimulates the various prevailing mechanisms of wear. At a low speed, abrasive-mechanical wear prevails, followed by adhesive wear, which is replaced by diffusion wear with a further increase in speed [1-4]. It is experimentally proved that the minimum wear is set at a speed $V_3^{(0)}$ corresponding to the transition state from adhesive to abrasive wear. However, depending on the formed attracting sets of deformation mixes in the interface zone of the tool and workpiece faces, there are bursts of forces that change instantaneous power values and, consequently, wear mechanisms. In other words, to reduce the cutting work on a given path when processing a group of parts, it is necessary to adjust the technological modes according to the criterion of minimizing the work on each discrete section Δt_2 in (10).

These features of evolutionary changes in the system determined the General algorithm for calculating the trajectories of the machine's Executive elements, on which the work of the forces under consideration is minimal. Therefore, the optimal trajectories are determined sequentially from the initial speed value $V_3^{(0)}$. At each step, the operation, wear increment, and change in parameters of the dynamic link formed by the cutting process are determined. In this case, the technological modes (cutting speed and feed rate) are adjusted so that when the condition (12) is met, the system's asymptotic stability or minimum operating conditions are provided.

3. Determining the optimal trajectory

The sequence of solving this problem is considered on the example of turning a part made of steel 08X15diameter D=80.0 mm. Turning with SANDVIK Coromant non-sharpenable plates made of GC2015 hard alloy, the shape of the plate is triangular type "W". Geometric parameters $\alpha = 2^0$; $\gamma = 6^0$, $\varphi = 90^0$. Processing without liquid coolant. In the dynamic subsystem of the tool, the axes of the ellipsoid of rigidity coincide with the axes of space $\Re^{(3)}$, since $\varphi = 90^0$; $m = \lfloor m_{s,k} \rfloor m_{s,s} = m_0$, $h = \lfloor h_{s,k} \rfloor h_{s,s} = h_0, m_{s,k} = h_{s,k} = 0, npu: s \neq k$. Parameters of the tool subsystem: $c_{1,0} = 3000 kg/mm$, $c_{2,0} = 1000 kg/mm$, $c_{3,0} = 600 kg/mm$; $h_0 = 5.0\ kgf/mm$; $m_0 = 0{,}025 kg \cdot f^2/mm$. Natural frequencies of the tool subsystem: $\Omega_1 = 350\ s^{-1}$, $\Omega_2 = 200\ s^{-1}$, $\Omega_3 = 120\ s^{-1}$ parameters of dynamic communication are given in table 1. The parameters of the integral operator are given in table 2. An example of an evolutionary diagram of change X and Φ and the corresponding tool wear path along the back face are illustrated on fig. 2. All diagrams calculated for the case of processing with constant modes are unique, and they depend on the initial parameters and modes. At their initial stage (section "A-B"), there is a loss of stability, and, as a rule, the formation of a stable limit cycle, the scope of fluctuations of which monotonously decreases.

Table 1. Parameters of dynamic link of the cutting process

ρ, *kg/mm²*	ζ, *s/mm*	μ	ρ_0, *kg/mm*	$T^{(0)}$, *sec*	α_0, *rad*	k_α, *rad⁻¹*	k_T
500	10,0	0,5	2,0	0,01	0,166	10,0	0,3

Table 2. Parameters change of the integral equation when variation of cutting speed

Speed, *m/s*	T_1, *s*	T_2, *s*	β_1, mm(kgm)$^{-1}$	β_2	μ_V
0,6	680	2170	1,5*10-4	1,1	0,2
1,2	360	11250	1,3*10-4	1,1	0,2
1,8	250	780	0,9*10-4	1,2	0,3
2,4	146	420	1,3*10-4	1,3	0,4
3,0	95	235	0,8*10-4	1,3	0,4

Then, depending on the depth of cut at points "A,...F", bifurcations of attractive sets of deformations are observed. Examples of fragments of phase trajectories on the "A-B" section for $t_p^{(0)} = 3{,}0\ mm$ are shown in fig.3. They are characterized by a great variety. Until now, there have been no direct theoretical studies of the relationship between wear resistance and dynamics. However, there are experimental data confirming the dependence of wear on dynamic properties: the dependence of wear on stiffness; the diameter of the part under constant modes; technological modes; geometric parameters of the tool, etc. [1-4] Most of the explanations for this effect are found in various physical representations of interactions: mechanical "attrition", fatigue destruction of near-surface layers; molecular (adhesive, diffusion, etc.); thermodynamic; tribochemical, etc. However, when cutting, the primary source of all

IOP Conf. Series: Materials Science and Engineering **900** (2020) 012007 doi:10.1088/1757-899X/900/1/012007

interactions is mechanics, or more precisely, irreversible transformations in the contact area of the supplied energy. Moreover, the dynamics of mechanical interactions significantly enriches the quasi-static concepts accepted in cutting science [1-6]. If the cutting conditions under constant conditions provide adjustment of technological modes to ensure stable cutting, then theoretically it is possible to achieve a significant (two to three times) increase in the cutting path while reducing the processing performance by no more than 10%.

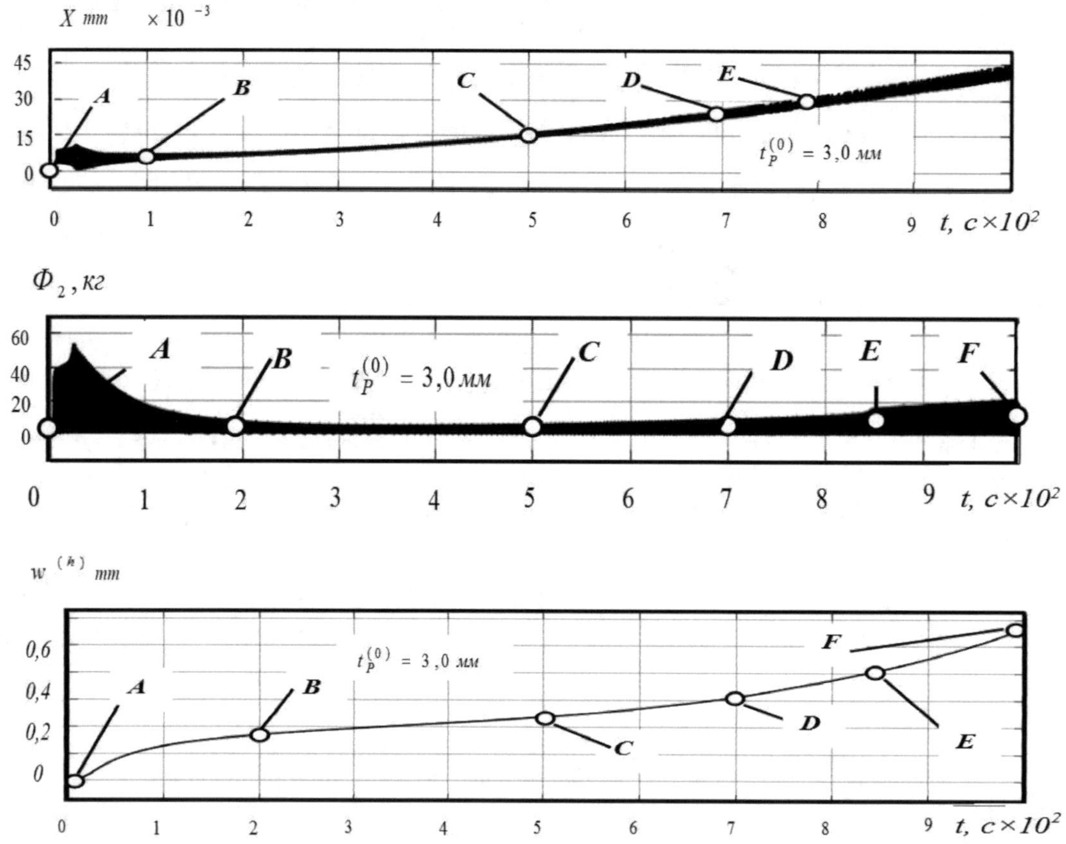

Figure 2. Wear $w^{(h)}$, change of deformational displacement trajectories in direction X_2 and forces Φ_2 are acting on rear face and corresponding to X_2

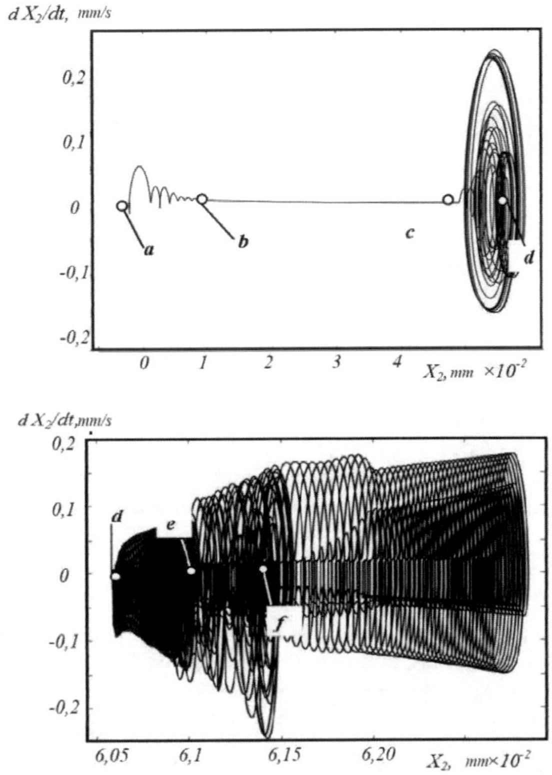

Figure 3. Phase trajectories with bifurcation point on section «A-B» (fig.2): «a-b» – transient process; «b-c» – stability movement; «c-d» – limit cycle; «d-e» – evolution of limit cycle; «e-f» – chaotic dynamic; «f-h» – limit cycle with new parameters

Figure 4 shows an example of this evolutionary trajectory. It reduces the cutting speed by increasing the chip pressure on the front surface of the tool. Finally, these results allow us to determine by the stability margin the processing conditions under which the minimum work is observed, which correspond to the conditions of the minimum wear intensity (Fig. 5).

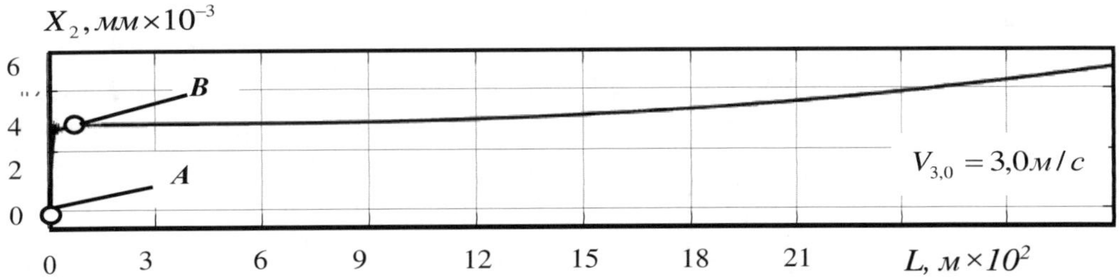

Figure 4. Change evolutionary trajectories of deformation displacement of the tool take into account cutting speed functional dependence on evolution phase.

In the illustrations, as the speed increases, there is initially an increase in the path, at which the wear reaches its critical value (0.8 mm). Then it is observed to decrease. The maximum stability margin corresponds to the speed range of changing self-excitation mechanisms. When processing a part with a smaller diameter, the extremum is shifted to the area of lower speeds. The dependencies shown in Fig. 5 correspond qualitatively to the known experimental data [4, p.219].

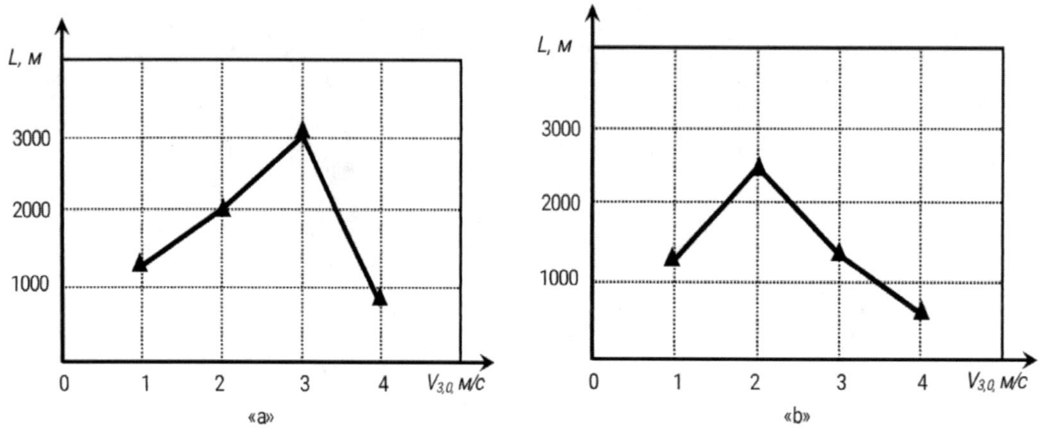

Figure 5. Results of digital experiment of measurement of tool path relative to detail with wear 0.8 *mm*: a – diameter spindle 50 *mm*; b – diameter spindle 20 *mm*.

4. Conclusion

A dynamic cutting system is a complex nonlinear transformer whose parameters change depending on the phase trajectory of the power of irreversible transformations of the input energy for the work performed. As a result, there is dynamic reconstruction takes place manifesting in changes in parameters and bifurcations of attracting sets of deformation displacements. The trajectory of work and power of irreversible transformations correspond it, and the evolution and properties of the dynamic system can be controlled by matching the controlled trajectories of Executive elements with evolutionarily changing parameters. As a result, the work of cutting forces on a given path significantly reduces. This determines the energy efficiency of cutting.

The study was funded by RFBR according to the research project № 19-08-00022.

References

[1] Ryzhkin A A 2019 *Synergetic of wear of tool materials at blade processing* (Rostov-on-Don: Don state technical University) p 289

[2] Starkov V K 2009 *The Physics and optimization of cutting of materials* (Moscow: Mashinostroenie) p 640.

[3] Loladze T N 1982 *Strength and wear-resistant of cutting tool* (Moscow: Mashinostroenie) p 320

[4] Makarov A.D. 1976 *Optimization of cutting processes* (Moscow: Mashinostroenie) p 278

[5] Vasin S A, Vereshchaka A S, Kushner V S 2001 *Metal Cutting: a thermomechanical approach to the system of interconnections in cutting* (Moscow: Bauman Moscow state technical University) p 447

[6] Ostafiev V A 1976 *Physical basis of metal cutting process* (Kiev: Kiev polytechnic institute) p 136

[7] Zakorotny V L, Gubanova A A, Lukyanov A D 2016 *Attracting sets when milling with end mills* STIN, 8, pp 27-33

[8] Zakorotny V L, Gvindjiliya V E 2019 *Dependence of tool wear and quality parameters of the surface formed by cutting on dynamic characteristics* Metal processing (technology, equipment, tools). 21(4) pp 31-46

[9] Bershadsky L I 1992 *Self-Organization of tribosystems and the concept of wear resistance* Friction and wear, 13(6) pp 1077-1094

[10] Kostetskiy B I *1986 Structural and energy adaptability of materials under friction* Friction and wear problems (Kiev), 6, pp 287-295

IOP Conf. Series: Materials Science and Engineering **900** (2020) 012007 doi:10.1088/1757-899X/900/1/012007

[11] Postnov V V, Shafikov A A 2012 *Development of an evolutionary model of cutting tool wear to control the processing process* Ufa: UGATU Bulletin, 11, 2 (29) pp 139 – 146.

[12] Ryzhkin A A, Shuchev K G, Moiseev D V, Vistoropskaya F A 2014 *On the issue of estimating the share of energy spent on wear in cutting conditions* Bulletin of DSTU, 3(78), vol 14 pp 152 – 163.

[13] Ryzhkin A A, Shuchev A I, Filipchuk M M 1985 *The Relationship of the characteristics of the thermal field of a friction pair with the intensity of wear* Friction and wear, vol 6, 1 pp 153-157

[14] Zakorotny V L, Fleck M B, Lukyanov A D, Voloshin D A *2004 Modeling of tool wear using integral operators* STIN, 3 pp 9-14

[15] Zakorotny V L, Gvindjiliya V E 2020 *Connection of self-organization of the dynamic cutting system with tool wear* University news. Applied nonlinear dynamics, vol 28, 1 pp. 46-61

[16] Zakorotny V L, Flek M B, Fam D T 2011 *Synergetic concept in the construction of precision control systems for manufacturing parts of complex geometric shapes* Bulletin of the Don state technical University, vol 11, 10 (61), pp 1785-1797

IOP Conf. Series: Materials Science and Engineering **900** (2020) 012008 doi:10.1088/1757-899X/900/1/012008

Improving the efficiency of greenhouses systems by developing algorithms for the synthesis of an optimal lighting spectrum based on the fundamental principles of agrophotonics

D Yu Donskoy*, E P Klyuchka and A D Lukyanov

Don State Technical University, Rostov-on-Don, 344000, Russia

*e-mail: dand22@bk.ru

Abstract. The question of creating an optimal microclimate for maximum realization of the genetically inherent potential in plants is far from being resolved. This is due to the multi-factor nature of the biotechnical system of greenhouses, the dependence on the number of controlled components of the microclimate and the consistency of their work: light, temperature, etc. Based on this task, we conducted initial research and developed additional lighting algorithms for greenhouse systems. Based on the work of previous years and existing scientific research, the most optimal method for optimizing artificial lighting and linking its intensity to the microclimatic parameters of the system has been determined.

1. Introduction

Algorithmization of management processes for various systems increases both their energy efficiency and quality for a wide range of indicators. The question of creating an optimal microclimate for maximum realization of the genetically inherent potential in plants is far from being resolved. This is due to the multi-factor nature of the biotechnical system of greenhouses, depending on the number of controlled components of the microclimate and the consistency of their operation: light mode, temperature mode, humidity mode, gas composition and nutrition mode. [1,2] in this paper, we consider the principle of algorithmization of the synthesis of the optimal lighting spectrum based on the fundamental principles of agrophotonics. To form the correct algorithms, we formed an experimental base, the objects of which were vegetable and green crops. In this work, the fundamental principles of agrophotonics are: the dependence of plants on lighting, its frequency and intensity, as well as the reaction of plants to a wide range of external influences. [2] the Fundamental complexity of this work is to determine the basic criteria of efficiency associated with the principles of agrophotonics and the resulting experimental base. [2]

2. Application of the least squares method for creating agricultural lighting systems with optimal illumination

Many tasks of agrophotonics require the synthesis of irradiation with a given spectrum to stimulate plant growth processes at different stages of development - from seed germination to maturity. The spectral characteristics of lighting significantly affect both the physical parameters of the developing plant and

IOP Conf. Series: Materials Science and Engineering **900** (2020) 012008 doi:10.1088/1757-899X/900/1/012008

its heredity, which leads to an increase in the viability of crops and their productivity. [1] the Use of this technology significantly increases the energy efficiency and modularity of additional lighting systems used in agro-cultivation of vegetable and green crops. However, after conducting a patent analysis and searching for scientific materials, we found that this technology is still at an early stage of development [1,8,9]. And the development of smart lamps is a little-studied area of agro-cultivation. Thus, we decided to consider the existing and developed methods of forming a given spectral density with a different set of LED LEDs present on the market.

We discussed the detailed formulation and analytical solution of the optimal spectrum approximation problem in the article "Simulation, identification and dynamic control of the luminaire of the synthesized spectrum" [7]. Our method is based on the least squares method, where the optimization function will be in the form (1):

$$J(k_1,k_2,...k_n)=\int_{\lambda_1}^{\lambda_2}\left(S(\lambda)-\sum_{i=1}^{n}k_iS_i(\lambda)\right)^2 d\lambda \xrightarrow[k_1...k_n]{} \min \quad (1)$$

where integration means that the approximation will not be based on the envelope, but on the total power of radiation and the spectral range from λ_1 to λ_2; $S(\lambda)$ – the functional or tabular spectral density to which the approximation is made; S_i – the functional or tabular spectral density of the LEDs used [7]

Thus, the use of the least squares method provides high accuracy of approximation, allows you to get an approximation with the smallest error when using LEDs with an abnormal view of the spectrum, such as full-spectrum LEDs, white, mixed type. The first graph in Figure 1 shows all the led modules we use. An example of approximating the spectral densities of discrete light sources can be seen in Figure 2, where graphs of the relative intensity from the radiation wavelength are shown.

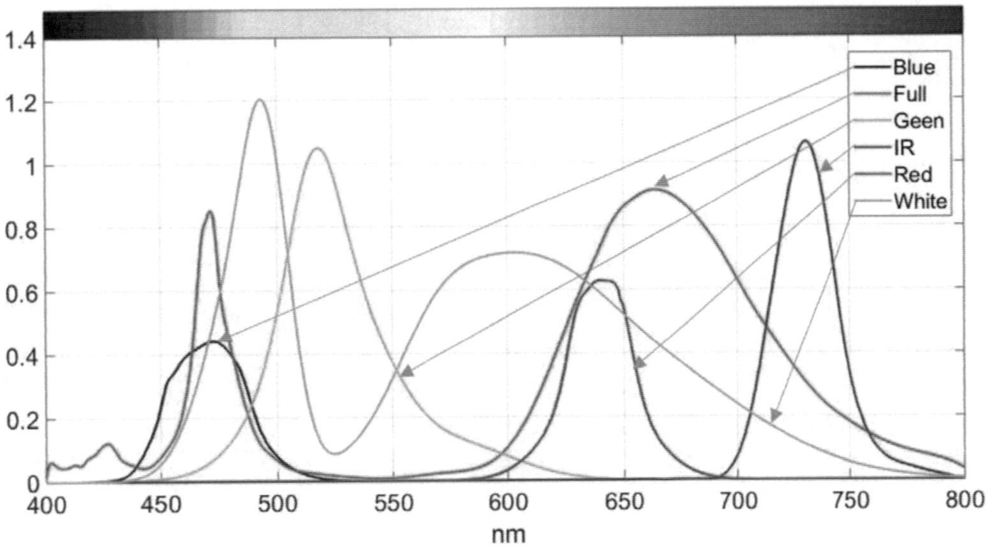

Figure 1. Spectral densities of 4 types of LEDs

IOP Conf. Series: Materials Science and Engineering **900** (2020) 012008 doi:10.1088/1757-899X/900/1/012008

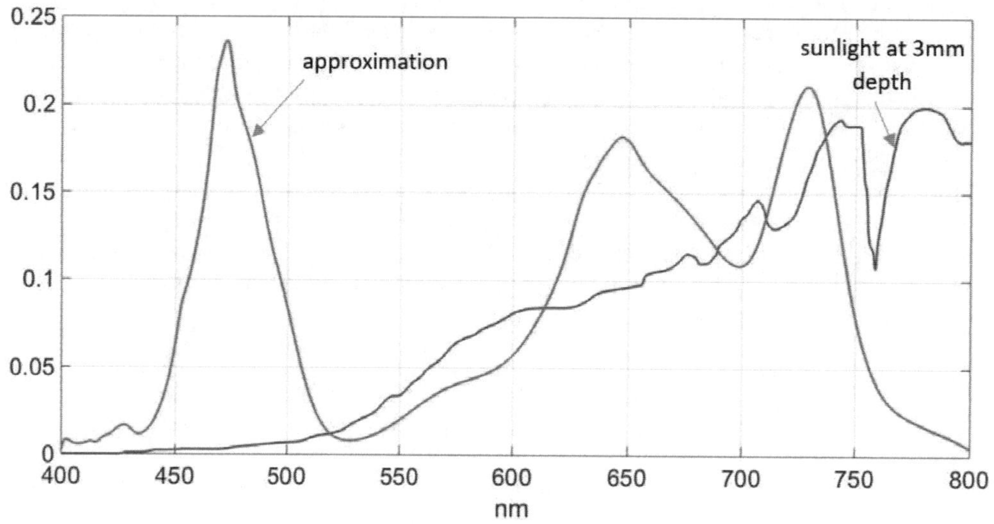

Figure 2. Approximation spectral densities of 4 types of LEDs to sunlight, modified when passing through 3 mm of soil. [6]

To reduce the intensity of the spectra at a length from 400 to 550 nm, long-wave filters with a transmittance of 550 nm are used. Figure 3. [8, 10]

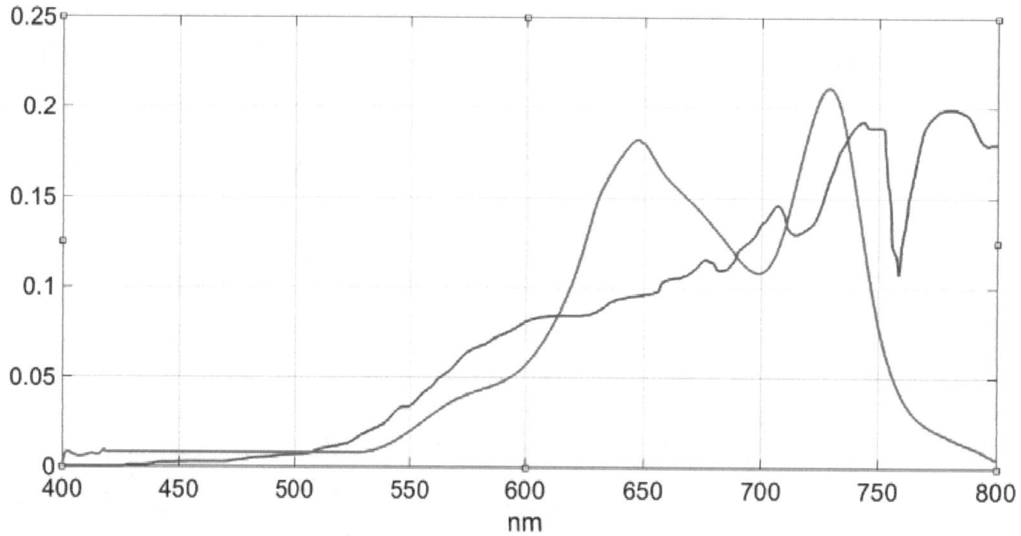

Figure 3. Approximation using a filter [6]

3. Adjustment of algorithms of stimulation of seeds on the microclimate

Having conducted a significant number of tests for tomato, lettuce and radish seeds, we have identified the most favourable microclimate conditions at the moment. Seeds are a dynamically developing biological object that requires equally dynamically changing conditions.

Despite the fact that it is not yet possible to analyse the issues of microclimate modelling comprehensively due to the huge number of factors and state variables that occur in a biological object, we have identified the main parameters of the installation - biological object system based on the research database, which at the moment can be regulated or controlled in any way. and determined their periodic fluctuations [1].

Based on the results obtained from the tests, it was determined that small temperature fluctuations (within 2 – 3 °C) do not significantly affect the culture of radish, lettuce and tomato seeds. The method of forming the temperature regime can be considered on the example of an algorithm for maintaining a microclimate for lettuce seeds.

The entire preparation period, for example, for lettuce seeds, the most favourable temperature is formed at 6-15 °C. The temperature changes from the lowest temperature point to the highest and Vice versa occurs in accordance with the period of light stimulation that occurs at the time of seed preparation. The stimulation period is selected for each crop and has a range of effective action, for example, for lettuce seeds, the smallest period is 18 hours. During this time, the light intensity increases to the maximum value and fades out according to the normal law. An example of temperature and light correlation for lettuce seeds is shown in Figure 4. This schedule is based on intermediate tests on the use of variable light stimulation when preparing seeds for germination.

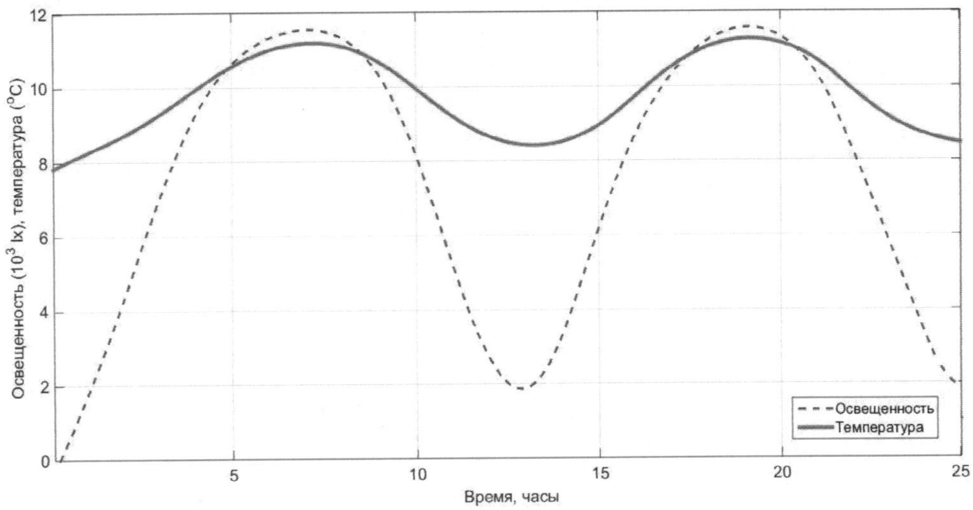

Figure 4. Correlation of temperature and light for lettuce seeds over 25 hours (~2 lighting periods)

Thus, the correction of the temperature regime is carried out by a sinusoidal law with the formation of a favourable temperature regime that correlates with the illumination with an approximately equal period.

Temperature changes that correlate with illumination are individually selected for each seed culture we studied.

4. Conclusion

The algorithmization of pre-lighting processes proposed in this paper determines the possibility of energy-efficient and all-season closed-type agrocultivation. Thus, for the further development of light-stimulation algorithms, it is necessary to constantly develop the experimental base. The application of our methods increases the percentage of germination and the rate of germination of vegetable and green crops while maintaining their high level of viability. Potentially, the use of the synthesized spectrum, taking into account the correlation with the microclimate, provides an increase in germination by 15-20 %. [6] Statistics are based on studies of the reaction of plants to synthesized lighting, which were considered in the article "study of the influence of the main spectra of discrete light sources on the seeds of greenhouse crops". [11]

Acknowledgments
Thank you to the Don State Technical University for their help in conducting experiments and to the University of Kragujevac Serbia for their help in decoding them.

References
[1] E. P. Klyuchka *Science and the world, Publishing house* Scientific review (Volgograd), ISSN: 2308-4804
[2] Degtyarev A. A., Klyuchka E. P. 2015. International student scientific Bulletin. **2-3**. URL: http://www.eduherald.ru/ru/article/view?id=12301 (accessed: 03.01.2019)
[3] Sano H. 2010. Inheritance of acquired traits in plants: reinstatement of Lamarck. Plant Signal Behav. **5(4)**: 346348. doi:10.4161/psb.5.4.10803
[4] Abeuov S. K., Altybayeva A. K. Seed Science: Educational and methodological guide for agronomic specialties of higher educational institutions. Pavlodar: Kereku, 2016.
[5] GOST 12038-84 "Seeds of agricultural crops. Methods for determining germination". Moscow. 2011.
[6] Babich S. V. Improving the methodology for assessing the germination of vegetable seeds: autoref. Diss. on the Internet. scientist. step. Cand. village-farm. Science. 2009.
[7] D. Yu. Donskoy, A.D. Lukyanov, M. A. Vernesi 2018. *MATEC Web of Conferences* **226**. Article Number 02030. URL: https://doi.org/10.1051/matecconf/201822602030
[8] D. Bliss, H. Smith. 2006 *Plant Cell and Environment* **8(7):** 475–483. DOI: 10.1111/j.1365-3040.1985.tb01683.x
[9] Kim, Hyeon-Hye & Wheeler, Ray & Sager, John & Yorio, Neil & Goins, Gregory. 2005. Light-emitting diodes as an illumination source for plants: A review of research at Kennedy Space Center. Habitation (Elmsford, N.Y.). **10.** 71-8. 10.3727/154296605774791232.
[10] Dutta Gupta, Snehasish & Agarwal, Avinash. 2017. Artificial Lighting System for Plant Growth and Development: Chronological Advancement, Working Principles, and Comparative Assessment. 10.1007/978-981-10-5807-3_1
[11] D Yu Donskoy, M A Vernezi, O I Katin, K I Goryanina, S G Studennikova and M Petkovic. 2019. Study of the influence of the basic spectra of discrete light sources on the seeds of greenhouse cultures. 10.1088/1757-899X/680/1/012009

Analysis of factors affecting the strength of fixed bonds assembled using metal-polymer compositions

V E Zinoviev, P V Kharlamov, N V Zinoviev, R A Kornienko*

Rostov State Transport University, Rostov-on-Don, Russia

*e-mail: Kornienkoroman23@mail.ru

Abstract. The paper considers the relationship between the quality of the surface layer of the substrate and the strength of the fixed adhesive bonding of parts. A search was made for the dominant factors affecting the strength of fixed bonds. A block diagram of the mutual influence of the quality indicators of a fixed bond of parts using metal-polymer gap compensators is given. It has been proven that the quality of the surface layer of the substrate, in terms of aggregate and degree of importance of the factors, has the greatest dominant effect on the strength of a fixed bond assembled using metal-polymer composition

1. Introduction

In the last 50 years, significant successes have been achieved in the theory and technology of the recovery of stationary compounds by polymer compositions, which is associated, in particular, with the creation of devices and methods that make it possible to carry out various tests, from the simplest to break and peel to complex non-destructive methods for determining the static and dynamic strength of compounds. However, despite the obvious progress in the field of the creation and use of sealing polymer compositions, an accurate explanation and understanding of the relationship between the quality of the surface layer of the substrate and the results of adhesion have not been achieved.

There are numerous theories of adhesion that explain the whole complex of adhesive phenomena in different ways [1-6]. There is currently no suitable mathematical or experimental apparatus for testing some of them. In most cases, these theories consider adhesion as the result of the interaction of molecules [1, 2], between which physical, chemical and intermolecular forces can act. To this should be added the interactions at the supramolecular level. Currently, the following adhesion theories are most used: molecular (adsorption), electrostatic, diffusion, rheological and chemical [1].

2. Mechanism of adhesion bonds formation in metal-polymer systems

In our opinion, the adhesion of polymer sealing compounds to the surface of the substrate (most often metal and plastic), as a rule, is caused, firstly, by micro-rheological processes occurring at the interface, and, secondly, diffusion. The micro-rheological mechanism of the formation of adhesive bonds is determined by the state of the surface of the substrates, their supramolecular structure, as well as the number of microdefects on their surface.

A number of authors [2, 5, 7] note that the mechanism of formation of adhesive bonds in metal-polymer systems is associated with the nature of bonds in the contact zone and their number. In turn, the nature of the bonds at the phase boundary (in the zone of formation of the adhesive contact) is determined by the structure of the adhesive and the substrate, which determines the corresponding nature of the interaction: chemical, intermolecular, electrostatic, etc. However, regardless of the nature of the bonds, their number is determined by the actual contact area of contacting surfaces. An increase in the number of bonds, due to an increase in the actual contact area, is accompanied by an increase in adhesive strength.

When studying the nature of the destruction of adhesive bonds, it was found [2,5] that it never occurred along the interface, but is cohesive. However, this theory does not answer the main question: as a result of which there is a bond and how various forces (van der Waals, dipole-quadrupole forces, etc.) affect its strength.

However, it is necessary to distinguish between the forces causing adhesion and the bond strength itself. In order to understand the essence of the process of sealing the gap with a polymer layer in the fixed conjugation of parts, it is necessary to understand quite well the effect of intermolecular forces that affect adhesion and cohesion.

In our opinion, the surface quality of the substrate and the degree of wetting by the adhesive composition have the greatest impact on the quality of adhesion and the strength of the sealing layer. Moreover, there are a number of other factors that, when combined, can sufficiently affect the strength of the product and its operational properties, which also must be taken into account when choosing the assembly technology.

Existing adhesion theories do not provide a satisfactory answer to the question of the formation and properties of fixed joints of parts assembled using a sealing polymer layer. When considering the processes of adhesion of the polymer to the substrate, we will try to determine the dominant factors that determine the quality of adhesion.

The experiments performed show that a qualitatively assembled immovable joint, as a rule, does not collapse at the substrate-polymer interface; therefore, at break it is impossible to speak of a pure destruction of the molecular forces that cause adhesion. It has been established that before the destruction of the system, partial or complete deformation of the entire compacting polymer layer occurs [1, 2, 8]. On the basis of which we can conclude that the work that is spent on the destruction of a fixed compound is spent not only on overcoming molecular forces.

The publication was carried out as part of a grant from Russian Railways for the development of scientific and pedagogical schools in the field of railway transport.

An inadequate bond is a bond the destruction of which occurs in the boundary layer. A strong boundary layer can be obtained by completely eliminating the weak layer, which is achieved in various ways. In some methods, the weak layer is not completely eliminated, but only pressed into the surface of the substrate. The appearance of weak boundary layers is caused by the presence of adsorbed gases and liquids, particles of solid substances on the surface, etc. A weak layer may also occur if the sealant or its formulation is incorrectly selected (polymer base, solvent, their ratio, additives, etc.), violation of the curing regimen or application technology.

One of the reasons for the occurrence of inadequate layers is that the polymer composition does not wet the substrate well, as a result of which air bubbles are formed during application, which remain after the adhesive has cured. Such bubbles at the phase boundary are usually very difficult to detect. It is easy to determine them only in the case of transparent adhesives or transparent substrates, for example, with lateral (cross) lighting [1, 9]. In case of the restoration of cylindrical fixed joints, the detection of such inclusions is extremely difficult. In addition, there are materials that are practically non-wettable, for example, polytetrafluoroethylene, polyethylene, polypropylene, etc. Comparison of the peel strength of polytetrafluoroethylene-steel and polyethylene-steel joints obtained by deposition showed that the strength of the second system is much lower. At the same time, the breaking load of the polytetrafluoroethylene-steel adhesive bonding is close to the strength of polytetrafluoroethylene itself. Based on this, Bikerman believes that in the polytetrafluoroethylene-steel system, an inadequate (loose) layer occurs due to air ingress, and in the case of polyethylene, the material itself has a major influence on the speed and force of destruction.

The degree of wetting can be quantified by the wetting angle, which is called the contact angle. It is impossible to determine the exact value of the contact angle due to difficulties stipulated by the surface properties, surface shape, and technological background of the substrate, therefore, an equilibrium contact angle is used. It is necessary to mention the wetting hysteresis - a phenomenon that is associated with the fact that a different volume of liquid can be on the same base of a drop, and, therefore, the contact angle will be different. The wetting hysteresis is manifested, as a rule, on an uneven surface, since on it the edge angle will be different than on perfectly smooth.

If there is no substance on the surface of the substrate that directly affects wetting, then the surface tension of the liquid polymer at the interface with the substrate affects the quality of the polymer layer. The relationship between the performance of adhesion and surface tension at the interface of the substrate-polymer system is expressed by the equation:

$$W_A = \gamma_{S^0} + \gamma_{LV^0} - \gamma_{SL}, \tag{1}$$

where W_A is the performance of adhesion;

γ_{S^0} is equilibrium surface tension at the "solid surface-air" boundary;

γ_{LV^0} is equilibrium surface tension between liquid and air;

γ_{SL} is surface tension at the boundary between solid and liquid.

The surface of a solid substance very quickly adsorbs liquid vapor, solvent, moisture, which are very difficult to remove before application or during polymer application. Therefore, it is generally accepted that the adhesion performance upon wetting will be less (by a fraction of the consumed) adsorption energy, and relation (1) is transformed as follows:

$$W_{A^X} = \gamma_{SV^0} + \gamma_{LV^0} - \gamma_{SL},$$

where γ_{SV^0} is surface tension or free surface energy of a solid substance at the equilibrium boundary with vapors.

Having analyzed the above formulas and reasoning, we can conclude that wetting a metal with a liquid polymer in the bond directly affects the adhesion quality and strength of the cured bonding layer. The work that will need to be done to destroy the resulting bond is determined by the surface energy of the system and is expressed by the following expression:

$$\Delta W = \gamma_2 + \gamma_3 - \gamma_{2,3},$$

where ΔW is the change in free surface energy per unit area that will occur when the compound breaks exactly along the substrate-polymer boundary;

γ_2 is polymer free surface energy (per surface unit);

γ_3 is free surface energy of the substrate (per unit surface);

$\gamma_{2,3}$ is free surface energy of the polymer-substrate boundary.

According to Bickerman, bonds are inadequate in the case when weak adhesion bonds or an insufficient number thereof arise. It is known that a layer of adsorbed gas and moisture is rapidly formed on the surface of materials, which has good adhesion to the substrate, but is mechanically much weaker. This layer, together with by-products from the polymer, forms a weak zone, which determines the low strength of the sealing layer. Sometimes these layers are called transitional.

The above considerations show that these indicators must be taken into account when developing the technology for assembly work and when choosing the modes of curing of the sealing polymer layer. These parameters are easily predicted in the conditions of modern assembly production and are assigned depending on the parameters of the part, its shape and size.

We believe that the factor determining the strength of the adhesive joint is the surface quality of the substrate. The study of the relationship between the polymer wettability of the substrate and the quality of the surface layer was often carried out from the point of view of chemical processes, and did not widely address the issues of surface preparation for assembly, which relates to the field of engineering technology. The available information is scattered and not generalized, comprehensive recommendations have not yet been developed, and optimal values of the surface layer quality indicators and their effect on wetting hysteresis have not been obtained.

When assessing surface quality, periodic changes along the surface and perpendicular to it can be noted. Along the surface, these can be geometric changes, anisotropy of physical and physico-mechanical properties (regions with different densities, crystal structures, stiffness, etc.), changes due to impurities and various external influences. Perpendicular to the surface, surface adsorbed and surface-oxidized or chemically modified layers appear, as well as changes in the composition, density, stiffness of the material, electrical conductivity, porosity.

Dependencies that allow controlling the technological process of compound formation when applying a polymer melt to a substrate containing microroughnesses are presented in the following form. The kinetics of leakage of the adhesive melt in the microroughness of the substrate surface determines the process of formation of the nominal contact area $S_н$:

$$S_н \cong \sqrt{\frac{P \cdot t}{\eta}},$$

where P is gluing pressure;

t is pressure contact time P;

η is effective viscosity of the polymer composition.

Each contact surface unit has n_i units of bonds of a certain type (chemical, intermolecular, electrical, etc.). Each type of bond is characterized by its energy value U_i, which should be spent to break the connection. In accordance with this, the adhesive strength can be represented as:

$$A \cong K' \sqrt{\frac{P \cdot t}{\eta}} \sum_{1}^{i=k} U_i \cdot n_i . \quad (2)$$

Coefficient K' should include parameters reflecting the dependence of A on the conditions for the destruction of gluing:

$$K' = k \cdot v \cdot e^{\frac{U_p}{R \cdot T_p}},$$

where v is gluing destruction rate;

T_p is temperature at which destruction takes place;

U_p is energy spent on an elementary act of breaking bonds;

k is coefficient taking into account the material of the parts to be joined.

The important factor affecting the adhesion strength is the gradient of mechanical properties normal to the surface of the substrate $\frac{\partial \Theta}{\partial z}$. It is obvious that the gradient of the mechanical properties of the third body will directly affect the coefficient of friction between the adhesive and the substrate, which, in turn, will affect the adhesive strength. According to the research of I.V. Kragelsky, deep tearing (destruction of bond) occurs when the threshold of external friction is violated due to the appearance of a negative gradient of mechanical properties along the depth of the friction surface or due to too large relative penetration. Such destruction is in the nature of tearing or puncturing the material not at the place of gluing, but inside one of the bodies.

Taking into account possible residual stresses in gluing $\sigma_{ост}$, contributing to delamination of the sealing polymer layer and the conversion of formula 2, the quantitative dependence of the adhesive strength on technological factors in the case of micro-rheological formation mechanism can be represented as:

$$A \cong k \cdot v \sqrt{\frac{P \cdot t}{\eta}} \cdot \left(\sum_{1}^{i=k} U_i \cdot n_i \right) e^{\frac{U_p}{R \cdot T_p}} - f(\sigma_{\text{ост}}) - \left(\frac{\partial \Theta}{\partial z} \right) .$$

It has been experimentally proved that surfaces machined to absolute smoothness, applied to each other and pressed by a load *P*, can be torn off by a force *F*, which is greater than *P*, i.e. $F/P > 1$. If this operation is carried out in a "high vacuum", which will cause the desorption of weak layers, then $F/P \cong 10^5$. This refers to joining two solid materials without adhesive. Bonds using sealing polymer layers should be considered as a system consisting of five layers: substrate 1, transition (boundary) layer 1, polymer, transition layer 2, and substrate 2. A bond whose transition layers are so strong that destruction goes beyond them is considered to be correctly formed, solid bond. That is, the question again arises about the quality of the surface layer of the substrate and its preparation for assembly using a sealing polymer composition.

It has been established that the kinetics of the formation of the actual contact area is associated with the kinetics of the flow of the polymer melt into defects in the surface layer of the substrate, and also, creatively developing previously published works [5, 7], a method for calculating the true contact area based on the assessment of the defectiveness of the contacting surfaces was proposed. The total nominal contact area per unit area of the substrate can be calculated by the equation:

$$S_{\text{н}} = 2n_1 \frac{h_i}{\cos\alpha} + 2n_2 \frac{h_i}{\cos\alpha} n_2 \pi^2 d^2 \sqrt{\frac{P \cdot t}{\eta}} , \tag{3}$$

where n_1 is the number of grooves (depressions) per unit surface of the substrate; h_i is maximum depth of inflow of the polymer composition in the micro-roughness of the surface of a part; n_2 is the number of other defects (in this case, cylindrical pores are considered) per unit surface of the groove; d is средний диаметр дефекта или поры в поверхностном слое субстрата; α is half of the angle at the top of the groove section (depressions).

We slightly transform the presented formula to go to the generally accepted parameters of the surface roughness of the part. For example, the arithmetic mean deviation of the profile of parts within the base length is usually estimated by the following expression:

$$R_a = \frac{1}{n} \sum_{i=1}^{n} |h_i| .$$

In our opinion, the effect of other surface defects, such as pores, on the joint area will be less significant and, therefore, can be neglected. In addition, the surface roughness obtained by us during the mechanical processing of the substrate can be predicted with a greater degree of probability, which means that more reliable results can be obtained in calculating the theoretical strength of the compound. Transforming formula 3, taking into account the above considerations, we obtain:

$$S_{\text{н}} = \frac{2nR_a}{\cos\alpha} \sqrt{\frac{P \cdot t}{\eta}} .$$

Since the performance of adhesion, regardless of the nature of the bonds, is proportional to the true contact surface, we obtain the following expression:

$$A = K \cdot E \left[\left(\frac{2nR_a}{\cos\alpha} \sqrt{\frac{P \cdot t}{\eta}} \right) - \left(\frac{\partial \Theta}{\partial z} \right) \right],$$

Where *A* is the adhesive strength; *K* is coefficient taking into account the nature of the stress distribution and the fluctuation mechanism of destruction; *E* is the energy of a single averaged adhesive bond multiplied by the number of adhesive bonds per unit of the true contact surface.

In the strength behavior of the adhesive bond, an important role is played by the character of the stress distribution in it, which is a function of the thickness and some other properties of the elements making up the bond. Therefore, the strength of the adhesive bond can also vary due to changes in the geometric or mechanical parameters of the surface layer. Although the nature of the processes that occur during surface treatment of substrates is different, all processing methods are aimed at increasing the surface energy of the substrate. Unfortunately, these factors are often neglected in explaining the behavior of the adhesive bond.

The assembly components of any modern technology have fixed bonds, which during operation should reliably and tightly fit to each other. These are shafts with disks, gears, rolling bearings rings pressed onto them, bearing shells pressed into the crankcases. Their immobility is ensured by friction between the pressed surfaces.

Under the influence of vibration and friction forces, the crystal lattice of pressed surface layers under cyclic tangential displacements becomes loose and collapses. The process of destruction is the abrasion of the surface without removing wear products. Detached metal particles oxidize rapidly. Oxides accumulate in contact, accelerating wear [1]. Methods of protecting mating from fretting corrosion should be aimed at preventing or drastically reducing the relative movement of contacting surfaces, corrosion-fatigue and related processes. Relative displacement occurs at the moment when the axial force from the external load exceeds the friction force in magnitude. In this case, it is necessary to reduce the axial load or increase the friction force.

It is possible to prevent relative displacement by transferring relative motion to the intermediate medium, for example, by applying a coating with a small elastic modulus to one or both contacting surfaces. Such material is polymers and, in particular, anaerobic materials, in which the elastic modulus is 200–500 times less than that of ferrous metals, and 70–160 times less than that of light alloys. For example, in anaerobic materials Anaterm-8K, Aantherm-6K, the elastic moduli are 60 MPa and 40 MPa, respectively, while in steels they are 150,000–200,000 MPa, and for aluminum alloys - 70,000–71,000 MPa. At the same time, the application of polymeric materials on the surface of ferrous metal parts protects them well from the effects of corrosion processes. [2].

One of the main tasks on which the performance of the bonds depends on is the determining the optimal method of surface preparation, taking into account the operating conditions of the adhesive structure. In the absence of such an approach under real conditions, the strength of the bonds may turn out to be below the acceptable level, which leads to premature failure of the bonds. At the moment, there is no single theory covering the whole variety of different methods of surface preparation. However, there is a set of recommendations obtained empirically that are applicable to various specific polymer – substrate systems and can achieve the desired result.

It must be borne in mind that the physicochemical properties of the surface layer of the substrate almost always completely do not correspond to the properties of the material and its mass. On the surface due to various influences, both physical and chemical, its own story is imprinted, which is individual for each subject and cannot be reliably evaluated as a factor affecting the bonding strength. There is only one way out - to reach an unchanged mass of material with a known composition and properties. Hence the simplest preparation methods based on the cleaning action of solvents, abrasives, which allows you to remove both surface contaminants and an altered surface layer.

More sophisticated processing methods are aimed at increasing adhesion by creating a surface layer having higher adhesion than the starting material. The prepared surface can be kept suitable for use for a long time by applying a protective layer that is removed in the future. Otherwise, it is recommended to apply a layer of glue on the prepared substrate as quickly as possible. Roughening mechanically allows you to get a surface with which the adhesive has better adhesion by increasing the contact area, and accordingly increases the mechanical adhesion. However, it must be borne in mind that the height of the obtained roughness must be deliberately less than the thickness of the adhesive layer.

If it is necessary to achieve a particularly high bonding strength, surface preparation should be performed especially carefully using more laborious surface treatment techniques. So, for example, in the case of gluing aluminum that has been cleaned only with a solvent, a rapid drop in strength is possible under the influence of climatic conditions. But the same acid-etched aluminum, assembled using the same polymer compounds and placed under the same climatic conditions, allows maintaining high strength of the fixed joint for many years. For the successful preparation of the surface of the part for subsequent assembly, it is necessary to more deeply analyze the basic quality indicators of the surface of the parts forming a fixed joint.

The surface quality of the substrate is determined by its geometric characteristics, physico-mechanical and physico-chemical properties of the surface layer [3, 4]. After machining and various types of action on the part, the surface layers of a solid have several layers interconnected with different forces. In the upper layers are molecules of adsorbed gas, water vapor and molecules of organic origin [3]. Depending on their concentration, they can form mono- or multi-molecular layers of homogeneous molecules randomly oriented relative to each other. Under this layer is a layer of molecules of organic origin, the molecular chains of which are parallel to the plane of the solid. This layer is bordered by a polycrystalline layer of polar molecules of organic origin. Further, layers of adsorbed gases and water vapors lie on the surface of the substrate, the layer of water vapor directly adjacent to the layer of oxides formed as a result of chemisorption of foreign molecules with the material of the surface layer of the substrate. The thickness of this layer is about 20–200 μm and depends on the medium in which the substrate was machined [3]. A strongly deformed riveted layer of substrate material is located under the oxide layer, which directly borders the main material.

3. Identification of dominant factors of contact surfaces quality and adhesion

Processing the surface layer of a part during manufacturing and other technological operations, no matter how perfect it may be, leads to a deviation of the surface profile from the ideal geometric shape specified by the drawing. Since the stationary conjugations of the "shaft-sleeve" and "body-ring" type were examined, in which the average diameter of the joint is comparable to or greater than the contact length, the following types of deviations, which can be divided into two groups according to related characteristics, are of interest. The first group includes barrel-shaped, concavity of cylindrical parts and surface taper. In the second group we combine non-circularity, ovality and cut.

When analyzing the first group, in relation to the aspects of the issue under study, one can draw the following conclusions. If the values of these errors are within normal limits, then they will not have a significant effect on the process of sealing the gap in the joint with a polymer composition. However, if their values (or the sum of the values) fall outside the tolerances, then a situation may arise in which part of the surface of the substrate will be excessively coated with the composition, while in some areas the surface will not be wetted at all or the composition will be applied in a minimal amount. Uneven distribution of the polymer layer over the surface of the substrate in the sealable gap will definitely cause a decrease in the strength of the joint and the system as a whole (Fig. 1) [3].

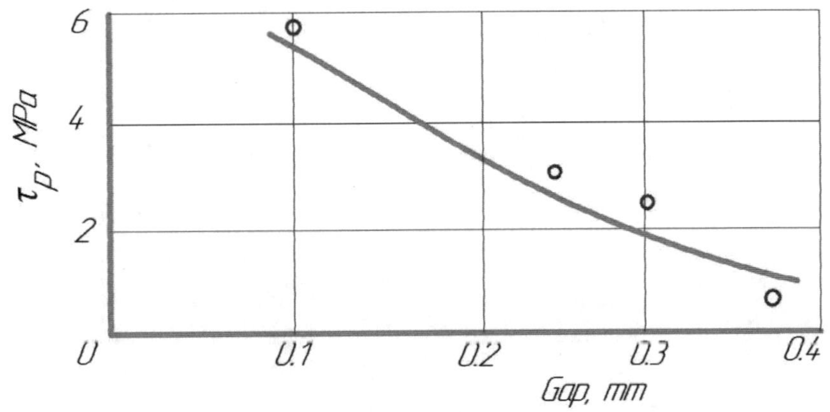

Figure 1. The effect of the gap between the joined surfaces on the strength of the sealing layer of the polymer composition (УГ-7)

The surface errors combined by us in the second group will also affect the quality and the process of forming the sealing polymer joint. As in the case considered by us above, exceeding deviations of permissible values can significantly worsen the process of wettability of the substrate surface by the polymer. As a result, a denser polymer seam uneven in thickness is formed only in a plane perpendicular to the axial lines of the cylindrical parts of the bond. However, in this case, it is possible to improve the uniform distribution of the composition on the surface of the substrate by turning the shaft or ring (alternating the direction of rotation), which will somewhat neutralize the harmful effects of these factors.

The surface roughness of the substrate has a direct effect on the strength of the assembled bond and the sealing joint. On the one hand, an increase in roughness contributes to an increase in the real contact area of the adhesive with the substrate in the compound (Table 1, Fig. 2), and on the other hand, it prevents the uniform and complete wetting of the surface of the substrate with the polymer composition. In addition, in the depressions on the surface of the substrate, if they are excessively deep, hard-to-remove inclusions of oils, dirt and air bubbles accumulate, which also have a negative effect on the strength of the joint, reducing it by 30-40% (Fig. 3) [4].

Table 1. Effect of contact area on the strength of cylindrical bonds (sliding fit, roughness R_a=1,25) [3]

Contact area, cm²	Axial thrust, N×m		Specific axial thrust, MPa	
	УГ-6	УГ-9	УГ-6	УГ-9
4,71	42	84	8,9	17,9
18,84	96	148	5,0	7,9
23,55	204	254	8,6	10,8
37,68	218	428	5,7	11,4
43,96	238	444	5,3	10,1
50,24	248	450	4,9	8,9
62,80	262	482	4,1	7,6
94,20	345	588	3,7	6,2

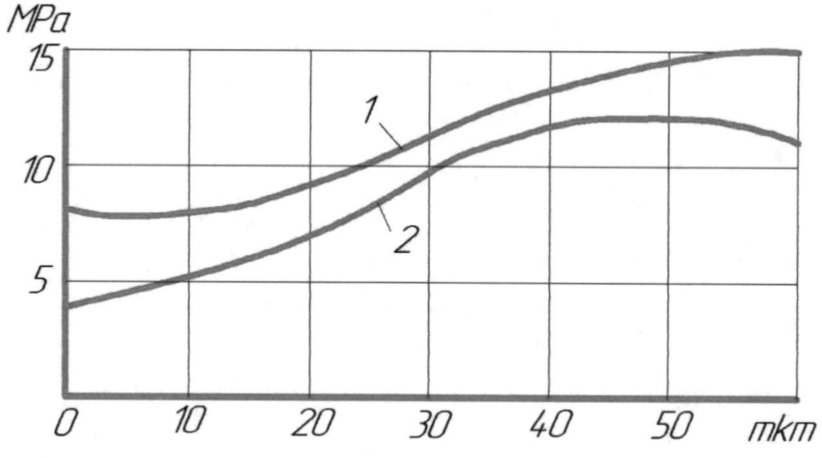

Figure 2. The dependence of the mechanical strength of the joint on the surface roughness: 1– material AN-6V (St. 45, clearance 0.25); 2 - material AN-6V (St. 45, clearance 0.25) [3]

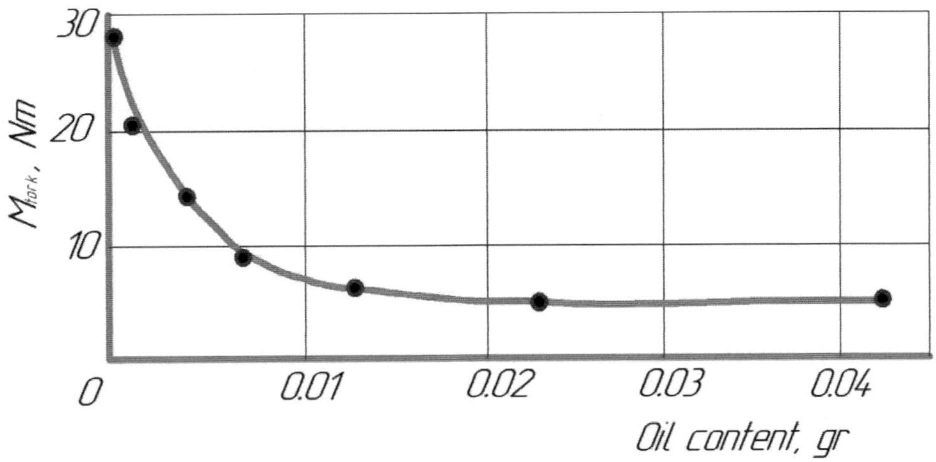

Figure 3. The dependence of the moment of turning the connection parts relative to each other on the oil content on the surface of these parts

The waviness of the surface of the substrate is a periodic micro-deviation from the geometric shape of the part, which are practically equal in size to the elevation and depression, with a wave step significantly exceeding its height. Distinguish between longitudinal and transverse undulation.

Waviness arises as a result of feed unevenness during turning and grinding, forced vibrations of the "machine-tool-tool-component" system, etc. The wave height is more important, from our point of view, criterion, since the following parameters depend on the height:

- the actual length of the bearing surface;
- volume of profile voids;
- specific pressure.

It is known that in press fits, the smaller the wave height and its step along the surface of the substrate, the larger the actual contact area. When using sealing polymer and metal-polymer compositions, when transitional and sliding fits are used, the role of undulation in ensuring the immobility and operability of part connections is somewhat reduced. However, in the literature there is no evidence of a positive or negative effect of this type of substrate surface errors on the assembly quality of the fixed bonds using sealing compounds. It seems interesting to us to identify which types of waviness, the dimensions of the wave heights and their lengths significantly reduce the strength of the connection, or increase it.

To identify the dominant factors of the quality of contact surfaces and adhesion, affecting the strength of fixed bonds, we combine the above material and reasoning in the form of a block diagram (Fig. 4).

Obviously, the indicators presented in the diagram can be divided into groups by importance in terms of ensuring the quality of the bond. Moreover, it seems difficult to vary all the indicators in the conditions of assembly or repair production. Therefore, it is necessary to determine a number of factors, the change of which is possible under the conditions of production of components and assemblies of various machines and equipment, and their change will allow to achieve the best quality of fixed bonds. After analyzing the compiled block diagram, it was concluded that the surface quality of the substrate in terms of the number of influential bonds with other indicators turns out to be the most dominant indicator of the compound. Therefore, in order to achieve good assembly results, it is necessary to propose methods for preparing the surface of the part for the upcoming operation.

In addition to the chemical and mechanical methods of surface preparation mentioned above, physical methods exist, such as: flame treatment, ion bombardment in vacuum (corona discharge) and electric discharge treatment. These processing methods are especially relevant for inert plastics such as polyethylene, polypropylene, polyethylene terephthalate, which simply cannot be effectively glued without such processing.

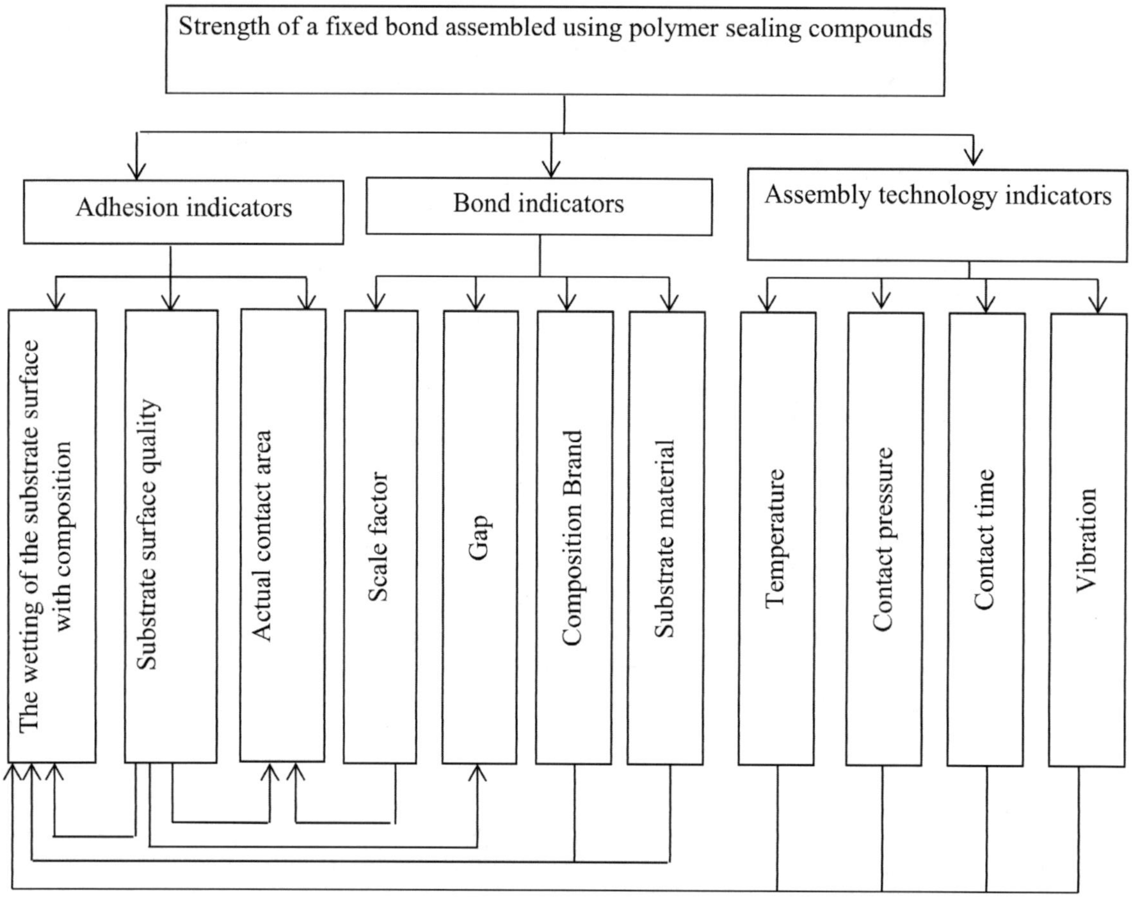

Figure 4. Factors and indicators affecting the quality of the fixed bond of parts using polymer sealing compounds

There are a number of reasons that cause a decrease in bond strength. The real chemical structure of the surface is quite complex, and information about its properties and the possibility of combining it with glue is often insufficient or completely absent. Therefore, to select the optimal surface treatment method, extensive experimental work should be carried out.

The essence of preparing the surface for gluing is to, using chemical, electrochemical, mechanical processes, the use of modifying additives, adhesive primers or other methods, change the nature of the surface of the substrate, make it more active when in contact with the adhesive to obtain the required strength. At the final choice of the surface preparation method, the design and technological features of the joint and the product as a whole, as well as the operating conditions, should be taken into account.

Thus, the choice of solvent type and degreasing method depends on the type of suspected surface contamination. After the degreasing process, regardless of where the part goes for gluing or for further processing, it is important to determine its effectiveness, the degree of surface cleanliness. In practice, it is determined by testing the continuity of the water film. This method allows you to give a relative assessment of changes in surface tension and determine the possibility of further technological operations.

In preparation for gluing metals, the following features exist. The freshly processed surface of metals with rare exceptions: gold, platinum, titanium and high-quality stainless steels is highly active, it is especially easily coated with an oxide film and is able to absorb airborne contaminants, dust and moisture. Therefore, we again recommend gluing as soon as possible after finishing the surface.

Metal surfaces are almost always contaminated with oils. Hence, a degreasing operation is mandatory, which can be done with water vapor, wiping the surface with a cloth dampened with a solvent (until traces of pollution no longer appear on a clean cloth). Treatment in solvent vapor is very effective. Degreasing should be thorough, in all cases it will not be superfluous to rinse the surface with a fresh portion of a clean solvent. Sometimes they are limited only degreasing. But when applying more complex methods of surface preparation, degreasing is in any case an obligatory first stage of processing. It is easy to check the effectiveness of degreasing by applying water to the treated metal. In the absence of oils or fats, a water film forms, flowing evenly, without droplets and tears.

After degreasing in any way, cleaning agents or detergent solutions should be used to remove dirt and other inorganic residues. Strong alkaline non-inhibited solutions are used for ferrous metals, titanium and some copper alloys. For aluminum and its alloys, when surface etching is to be avoided, inhibited solutions must be used. As inhibited alkaline solutions, synthetic detergent solutions can be used. Alkaline cleaning is usually accompanied by chemical or mechanical treatment, since alkali-treated surfaces are incompatible with most structural adhesives. Before bonding, the surface should be drained with hot air.

Mechanical, abrasive treatment is used when the metal surface bears a noticeable layer of corrosion or scale (steel, cast iron). Oxides are rapidly formed on aluminum and its alloys, in contrast to the inert metals mentioned above. Mechanical methods include those based on the abrasive effect of wire brushes, glass and sandpaper, shot blasting, which can remove unnecessary surface layer. While abrasive treatment roughens the surface and increases the contact area between the adhesive and the substrate, it should be noted that after this treatment the tendency for moisture or solvents to penetrate into the adhesive joint increases. This method may be completely unsuitable for parts that have undergone precise machining, where the final dimensions are strictly specified.

When treating surfaces by dry abrasive blasting, good results can only be achieved by using grains with good cutting edges, i.e. using materials with high hardness, such as aluminum, silica sand or carborundum. Round glass or metal balls can cause a metal surface to ripen and are therefore unsuitable. The nature and size of the grains are the important factor in shot blasting and sandblasting. For most metals and alloys, the grain size of the abrasive and the processing methodology were empirically determined. In addition, the grains of abrasives themselves should not carry pollutants that can nullify all the work. A final rinse with a solvent removes abrasive residues, the presence of which on the adhesive-substrate section will weaken adhesion. If it is possible to use blown moistened abrasives, you need to know that this method is preferable, since it has a wider range of abrasive materials that can be used. If water is used as a humidifier, the use is limited only to non-corrosive metals or corrosion inhibitors must be added to water.

Chemical and electrochemical methods of surface treatment, when compared with mechanical ones, are more effective, cheaper and give good reproducibility of the results. Besides the fact that they have a cleaning effect, these types of processing often provide the formation of a chemical layer on the surface, contributing to the long-term preservation of the strength of the adhesive bond. The process called etching involves immersing the substrate in solutions of acids or alkalis (from dilute to strong) at various temperatures. Acids or alkalis destroy the surface layer of oxides faster than the base metal. If necessary, inhibitors must be added to the acid solution, which increase the difference in the dissolution rate of the metal and oxide. This is especially important for thick layers of scale, if for any reason the use of abrasive treatment is unacceptable.

After removing thick layers of oxides from the surface by etching, then the metal should be treated in weaker solutions, for example, in an "acid-dichromate" bath, which leads to the formation of a thin oxide film with the necessary structure, firmly bound to the main metal layer. The final rinsing and rinsing should be carried out in clean water, meaning the distilled water, which was stored under conditions excluding the ingress of foreign impurities, and unsaturated with atmospheric oxygen.

4. Conclusion

Analyzing the data and reasoning, as well as relying on the experience of other works [1-4], we can draw a number of conclusions:

- Achieving sufficient adhesion strength and operational reliability of the metal-polymer bond is possible provided that the maximum contact surface and active interaction of the adhesive and substrate at the interface are realized;
- the viscosity of the polymer composition should be at the time of application such that the filling time of the capillaries and cavities on the surface of the substrate is sufficiently small in comparison with the rise time of the adhesive viscosity during curing;
- good wetting of the substrate surface can be achieved if the contact angle tends to zero;
- surface wetting hysteresis is significantly affected by surface quality, the presence of contaminants and the viscosity of the polymer composition;
- the main indicators of the surface quality have a number of optimal values at which the maximum bonding area is achieved, but there is no reaction to wetting the surface of the polymer with a polymer;
- it is necessary to solve the issue of competent and technological removal of air and foreign inclusions from the polymer sealing layer;
- it seems interesting from the point of view of the mathematical apparatus to relate surface quality indicators, wetting hysteresis values and strength of the assembled bond.
- as a rule, the list of sealing polymer compositions produced by the chemical industry is limited and stable. Changing the basic physical and mechanical characteristics of the composition under production conditions is not possible. The development of new compositions by manufacturers on order will inevitably cause a rise in the cost of production;
- changing the material of the substrate and the scale factor of the connection (the dimensions of the main parts that make up the connection) is also not advisable, since it is impossible without coordination with the developers of the project and will require a lot of time and effort;
- the wettability of the substrate surface with a polymer composition and the actual contact area directly depends on the quality of the surface layer of the substrate, the viscosity of the polymer composition and the pressure in the bond;
- the surface quality of the substrate in terms of the number of influential bonds with other indicators is the most dominant indicator of the connection;
- the main indicators of the quality of the surface layer of the substrate, taking into account the large number of mechanical and other types of surface preparation, can easily vary;
- a change in the technology for the production or repair of fixed bonds can be carried out in mass production without significant costs and higher production costs..

Researches are conducted within realization of a grant of JSC "Russian Railways" on development of scientific and pedagogical schools in the field of railway transport.

References

[1] Boyko N.I. Resource-Saving Technologies for Vehicle Repair with Metal-Polymer Compositions: Monograph / N.I. Boyko, V.E.Zinoviev. M. : Marshrut, 2003. 250.
[2] Vasiliev A.S. Technological Fundamentals of Machine Quality Management / A.S. Vasiliev [et. al.]. – M. : Mashinostroenie, 2003. 256.
[3] Suslov, A.G. Scientific Foundations of Engineering Technology / A.G. Suslov, A.M. Dalskiy. M. Mashinostroenie, 2002. 684.
[4] Kartashov E.M. Structural and Static Kinetics of Polymer Failure / E.M. Kartashov, B. Tsoy, V.V. Shevelev. M.: Khimiya, 2002. 736.

[5] Berlin A.A., Basin V.E. Polymer Adhesion Basics. M.: Khimiya, 1974. 391.
[6] Kovacic L. Bonding of Metals and Plastics. M.: Khimiya, 1985. 240.
[7] Van Krevelen D.V. Properties and Chemical Structure of Polymers. M.: Khimiya, 1976. 416.
[8] Kagan D.F., Gul V.E., Samarina L.D. Multilayer and Combined Film Materials. M.: Khimiya, 1989. 288.
[9] Lipatov Y.S. Physicochemical Fundamentals of Polymer Filling. M.: Khimiya, 1991. 260.
[10] Gul V.E., Kuleznev V.N. Structure and Mechanical Properties of Polymers. M.: Vyshaya Shkola, 1979. 352.
[11] Bartenev G.M. Strength and Fracture Mechanism of Polymers. M.: Khimiya, 1984. 280.
[12] New Methods for The Study of Polymers. Collection of Articles Edited by *Lipatov Y.S.*. – Kiev: Naukova Dumka, 1975. 200.
[13] Zinoviev V.E. Research of the Static Strength of Fixed Bonds Assembled Using Polymer Anaerobic Compounds / Zinoviev V.E., Khachkinayan A.E., Zinoviev N.V.// Izvestiya VolgSTU, January (No. 223) 2019 No. 1. 12-16.
[14] Zinoviev V.E. The Effect of Microcracks, Latent Defects and Residual Stresses of the Polymer Adhesive Layer On Its Destruction / Zinoviev V.E., Kharlamov P.V // Fundamentalnye Issledovaniya, 2015, No. 12. 37-42.

IOP Conf. Series: Materials Science and Engineering **900** (2020) 012010 doi:10.1088/1757-899X/900/1/012010

Investigation of the influence of heat treatment of aluminum alloy on its structure and properties

O V Kudryakov and A I Agibalova*

Don State Technical University, Rostov-on-Don, Russia

*e-mail: agibalova250894@mail.ru

Abstract. The paper considers the issue of improving the quality of aluminum alloys of the D16 type, based on the data performed at the Department of "Physical and applied materials science" within the framework of the contract on samples of the Belokalitvensky metallurgical combine. The influence of the composition and technology on the structure and mechanical properties was studied. A recommendation for optimizing the composition and structure was developed and based on the analysis of phase diagrams, the microstructures of the alloy were justified.

1. Introduction

This paper examines aluminum alloy D16 which is one of the most popular duralumin alloys in the shipbuilding, aviation and space industries. It has such characteristics as stable structure; high strength characteristics; 3 times lighter weight than steel products; good machinability on lathes and milling machines, second only to some other aluminum alloys. This will be considered in more detail based on the experimental part. Duralumin D16 related to aluminum alloys of the Al-Cu-Mg system, alloyed with manganese. Most of it is aluminum — up to 94.7%, the rest is accounted for by copper, magnesium and other impurities (Table 1). Manganese increases the corrosion resistance of the alloy and improves its mechanical properties, although it does not form common hardening phases with aluminum, only dispersed particles. [2].

Table 1. Chemical composition of D16 alloy according to GOST 4784-97 in bulk %

Alloy grade	Si	Fe	Cu	Mn	Mg	Cr	Zn	Ti	Ni	Other elements	Al
D16 (1160)	0,50	0,50	3,8-4,9	0,3-0,9	1,2-1,8	0,10	0,25	0,15	–	Ti+Zr≤0,2	The rest

D16 is a structural heat-strengthened and naturally-aged alloy in the billet, which is used in various areas of the national economy. It is also used for the manufacture of power elements of structures in aircraft: parts of the skin, frame, ribs, control rods, spars, etc. It is also used in the automotive industry for the manufacture of bodies, pipes and other fairly strong parts. D16 is used for manufacturing rivets with high shear strength [3].

In connection with the above, the purpose of this work was an examination of the effect of heat treatment on the microstructure and mechanical properties of aluminum alloy D16.

2. Material and experimental procedure

As mentioned above, aluminum alloy D16 was chosen as the research material. Experimental studies of the structure and mechanical properties were conducted in the laboratories of the Don State Technical University at the Department of Physical and Applied Materials Science.

The study used the original billet made of aluminum alloy D16. Two identical samples № 4 and № 5 (with the end mark "124") in a non-heat-treated state (figure 1) was taken, cut from a D16 alloy ingot (the composition is shown in Table 1).

Figure 1. General view and geometric dimensions of samples № 4 and № 5 made of D16 alloy

3. Results and discussion

The conditions for working with the sample should be set as follows. The holding time at the homogenization temperature should be from 4 to 18 hours. It is important to track the detection of inclusions. The structure of duralumin in the annealed state is an α-solid solution of copper substitution in aluminum, along the grain boundaries of which fine-dispersed $CuAl_2$ or other secretions are located.

Six control samples were made, each of which was continuously kept in the oven for 4, 6, 8, 12, 16 and 18 hours without extraction. Samples were loaded into the furnace at the same time. The experiment was carried out in a chamber furnace SNOL 6.7/1300 at a temperature of 485$\pm$5.0°C. The temperature was chosen based on the Al-Cu diagram (Figure 2). Heating to lower temperatures does not provide maximum mechanical properties while heating to higher temperatures causes burnout, i.e. oxidation and partial melting of the alloy along the grain boundary, which sharply reduces the strength and plasticity. Therefore, the temperature value was chosen correctly. Temperature control was carried out not only by a temperature indicator of the furnace but also by a verified chromel-alumel thermocouple.

After the specified time intervals, one of the samples was removed from the furnace, cooled in air, and underwent the necessary sample preparation operations to study the structure, chemical and phase composition.

The change in the structure of samples depending on the holding time in the furnace is shown in Figure 3 (ZEISS optical metallography, etching with reagent № 3, the composition of which looks like h_3po_4-75ml., h_2so_4-15ml. and hno_3-10 ml). It is important to clarify that the etching lasted 3-5 minutes.

The main characteristic of all presented microphotographs is that all phase selections are located along the grain boundaries. The grain size does not change with increasing exposure in the oven and is 40...70 microns (there are also larger crystallites). Figure 3 shows photos with different degrees of optical contrast, so the samples with an exposure time of 6, 12 and 16 hours (2, 4 and 5 rows of double photos) all phase selections are identical, and the remaining samples (4, 8 and 18 hours of exposure) these selections seem to belong to different phases. The analysis of the chemical and phase composition will be presented later.

According to the visual estimates of Figure 3, it is quite difficult to draw even qualitative conclusions about the phase formation kinetics as the sample holding time in the furnace increases.

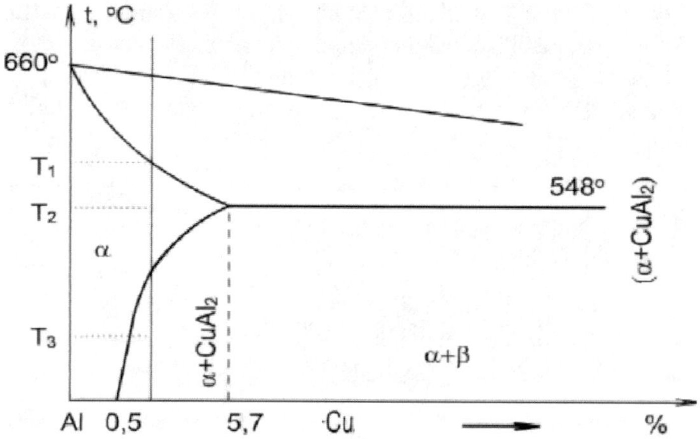

Figure 2. Phase Diagram of Al-Cu

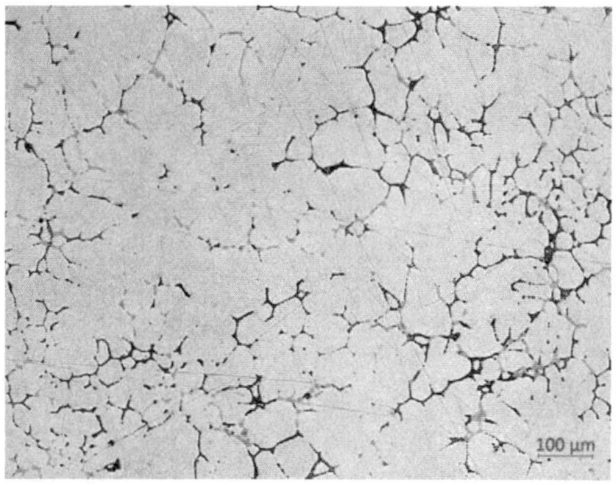

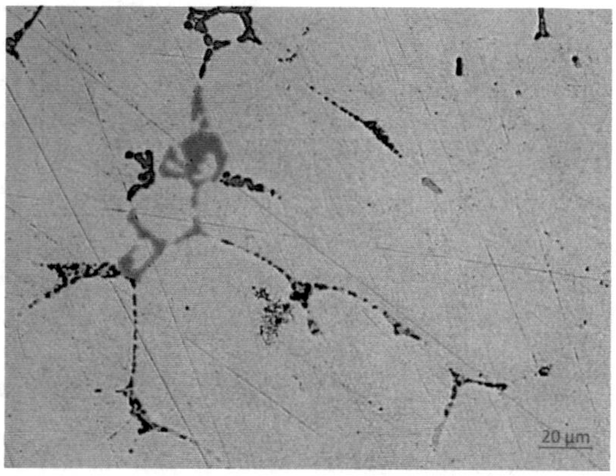

a) b)

c) d)

IOP Conf. Series: Materials Science and Engineering **900** (2020) 012010 doi:10.1088/1757-899X/900/1/012010

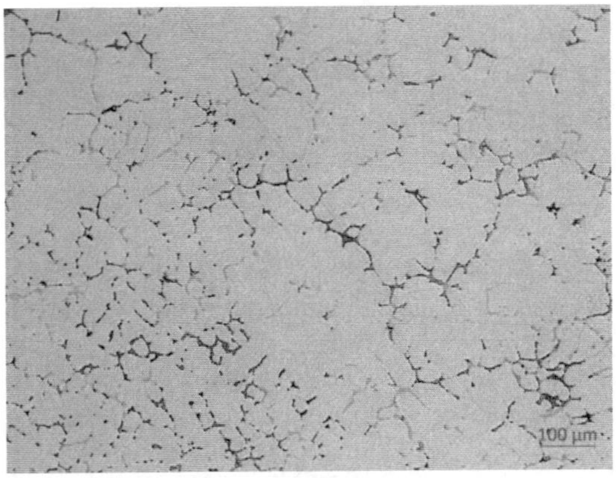

e)

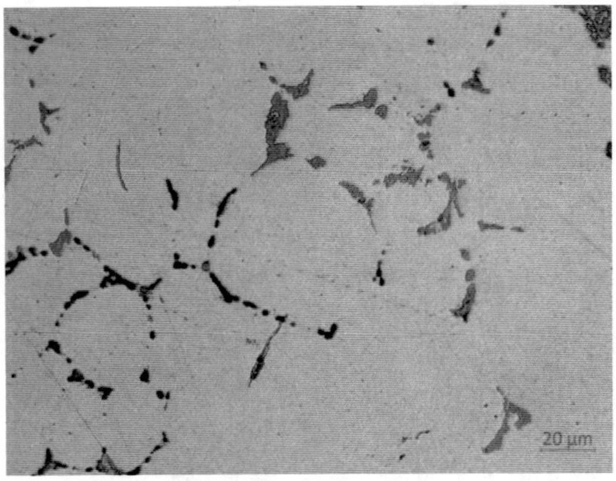

f)

g)

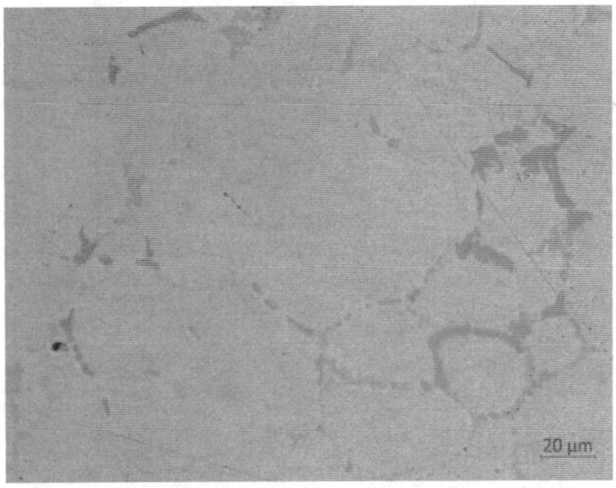

h)

i)

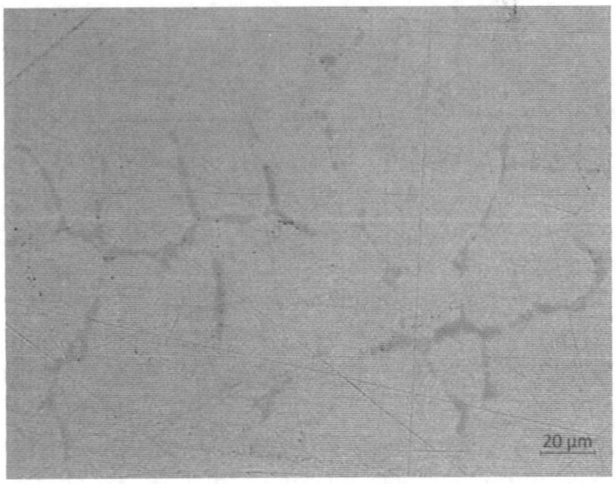

j)

IOP Conf. Series: Materials Science and Engineering **900** (2020) 012010 doi:10.1088/1757-899X/900/1/012010

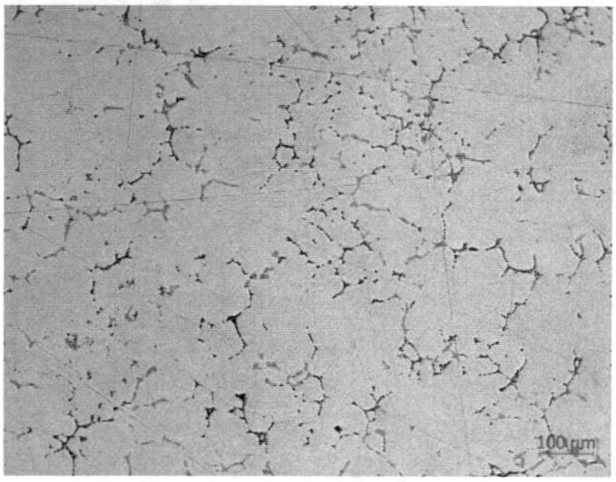

k)

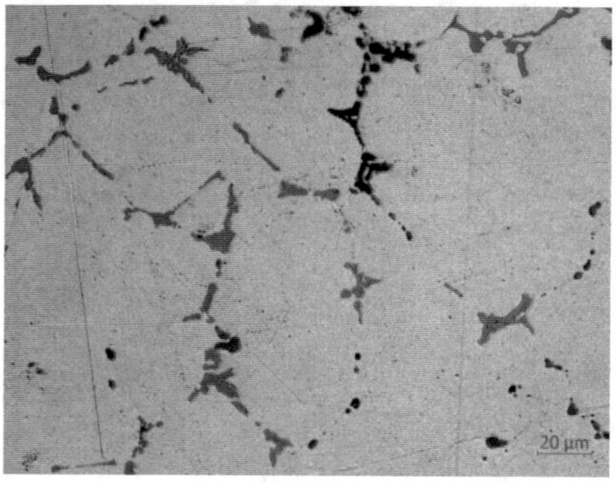

l)

Figure 3. The microstructure of 6 control samples at two magnifications, optical metallography: a, b – sample after 4 hours of exposure in the furnace; c, d – 6 hours; e, f – 8 hours; g, h – 12 hours; i, j – 16 hours; k, l – 18 hours

As you can see in Figure 3, as a result of heating, the main amount of $CuAl_2$ or other particles (the reducing phases will be described later) is dissolved in an α - solid solution. The D16 structure itself consists of a fine grain its shape and size are symmetrical. If you compare the microstructures, you can see the difference in the density of grains throughout the entire volume of the alloy and the presence of etching pits. This is just what the picture shows.

Therefore, to make a more accurate assessment, a statistical analysis of the total area of phase secretions in the samples was performed using the computer program "KOI – structure analysis". Figure 4 shows the measurement data for the amount of dark grain boundary phase (% of the total image area) in the sample structure after 12 hours of exposure in the oven (Figure 3, h) using the KOI program method.

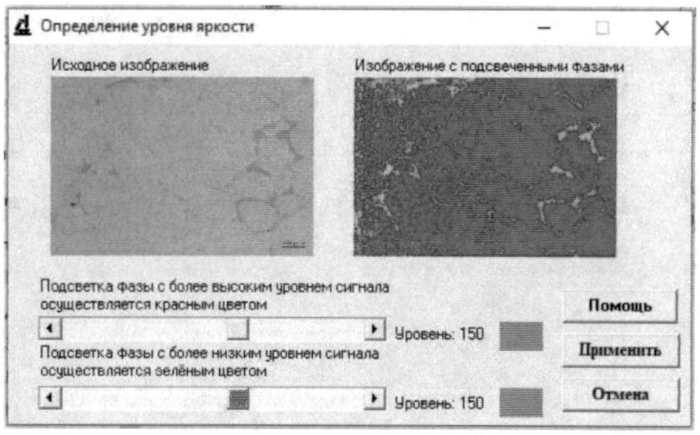

a.

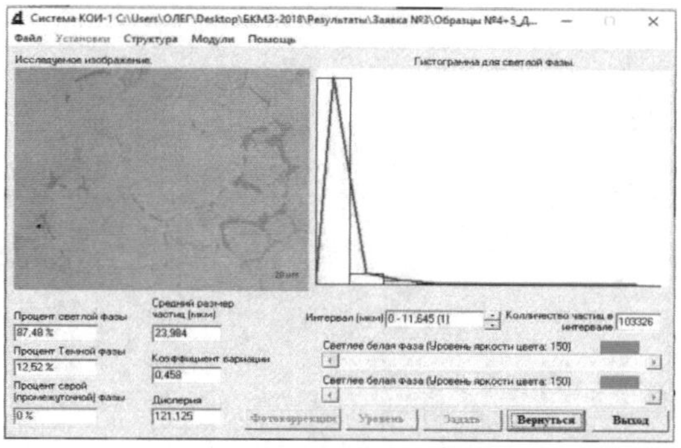

b.

Figure 4. General principles of operation in the image analysis program "KOI-structure analysis": a – window for determining the brightness level of the dark and light phases of the alloy (for the shown image, the boundary level is 150 out of 255); b – results of computer analysis – 12.52% of the total area is occupied by the dark phase

Three randomly selected optical microstructure images of each of the six control samples were analyzed using the given KOI method. The results of statistical processing have given a stable growth trend of phase grain boundary discharge, which is graphically shown in Figure 5. The growth rate of the grain boundary phase area in accordance with the experimental dependence of Figure 5 is approximately

0.7% per hour. This rate of growth is quite problematic to catch in the visual analysis of images (Figure 3). Moreover, the spread of the values of the area of phase allocations on a single micro shliff, calculated in the KOI program, sometimes reached up to 40%. The difficulties of visual analysis are also related to the fact that the growth of the dark phase occurred only along the grain boundaries. In the studied samples, phase separation in the grain volume was not observed.

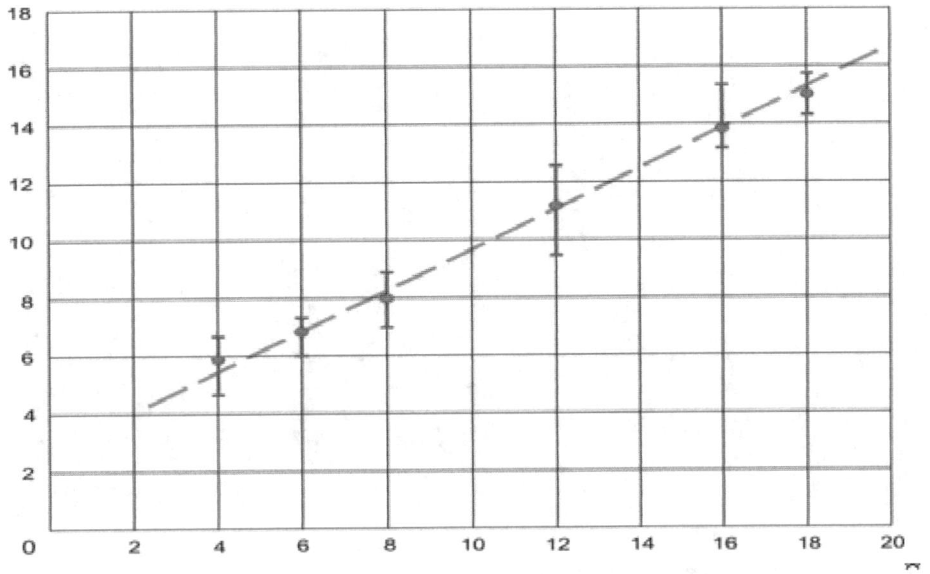

Figure 5. Experimental dependence of the change in % of the micro shliff area occupied by the grain boundary phase on the holding time in the furnace at a temperature of 485±5.0°C for D16 alloy

The elemental composition of inclusions in the test samples was studied by energy-dispersion analysis (EDAX) (performed using an energy-dispersion spectrometer based on the gas-free x-Mach detector of the company "Oxford instruments", mounted on a transmission electron microscope JEM-2100). Its results in the form of point probing (spectra) and a color map of the distribution of elements are shown in Figure 6 for a sample kept in a furnace during homogenization for 18 hours.

The results obtained for the remaining samples are not given since the morphology and composition of inclusions in them remained constant (only the volume fraction of inclusions changed during aging). In the dispersion-hardening (thermally strengthened) alloy D16, the following hardening phases may exist [1, 3]: $CuAl_2$, $MgCuAl_2$ and $Cu_2Mn_3Al_20$. According to Figure 6, there are two types of phases among the grain boundary secretions – copper-based and iron-based. Both of these phases are located in the intergrain space, adjacent to each other, but localized separately. This is clearly visible on the color map. Analysis of the obtained data allows to conclude that the composition of these phases is $MgCuAl_2$ and FeMnSi. Moreover, the dependence shown in Figure 5 on the growth of the grain boundary phase as it is aged in the furnace refers to the $MgCuAl_2$ phase, which "sprouts" along the grain boundaries due to the diffusion of alloying elements of the alloy from the grain volume. The number of FeMnSi phases does not depend on the duration of the homogenization process. Since Fe and Si are not part of the D16 alloy, the data obtained suggest that FeMnSi particles are exogenous inclusions (particles of flux, slag, modifier, lining, etc.) that were present in the melt.

Since FeMnSi particles are not isomorphic to the α - solid solution, they are displaced to the boundaries of growing crystallite–phase crystallites during crystallization. If the FeMnSi particles are isomorphic with $MgCuAl_2$, they can serve as a substrate for the crystallization and growth of this phase along the grain boundaries of the α - solid solution during the homogenization process. This mechanism of growth of the $MgCuAl_2$ phase is fundamentally different from the mechanism of separation of dispersed particles of the $CuAl_2$ phase in the volume of grain α –solid solution during aging. Thus, it is shown that the phase formation in the presented samples № 4-5 of the D16 alloy during homogenization occurs by forming the $MgCuAl_2$ phase by the mechanism of heterogeneous (on FeMnSi inclusions) grain boundary growth in accordance with the kinetics shown in Figure 5.

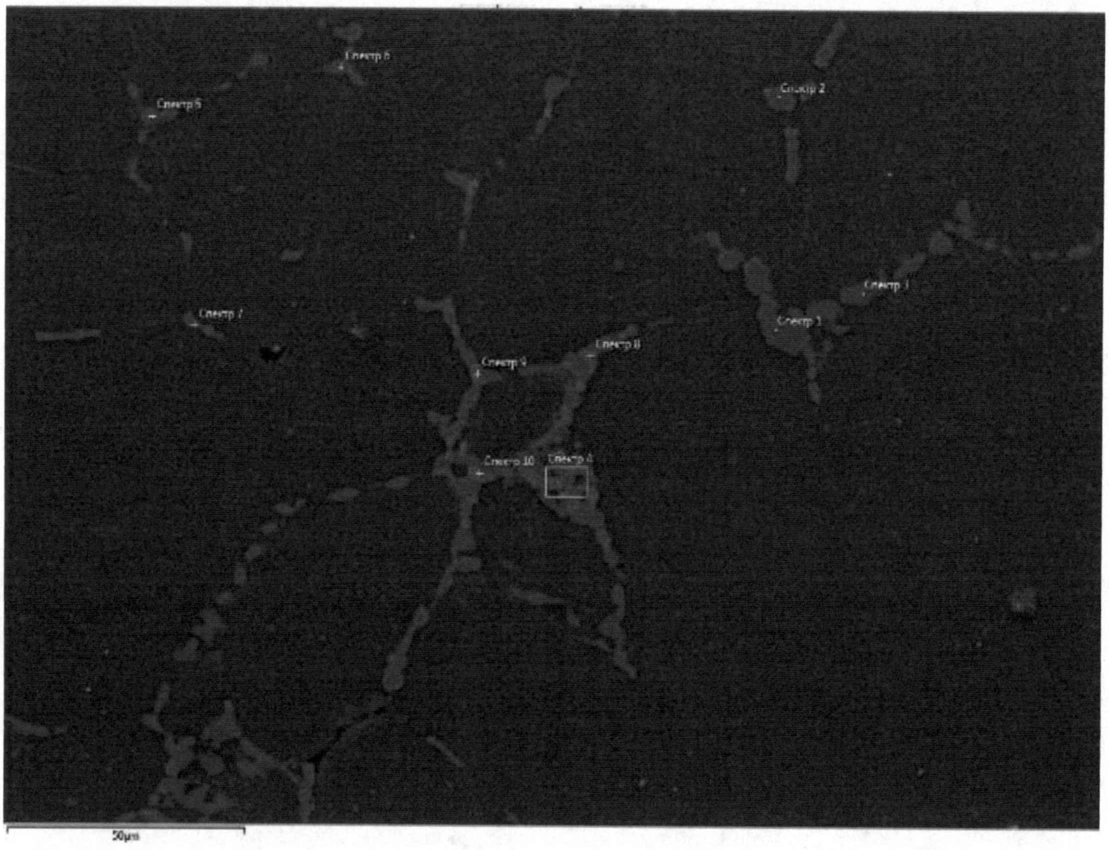

The name of the spectrum	Mg	Al	Si	Mn	Fe	Ni	Cu	The amount (% at)
Spectrum 1		71,33	3,86	6,17	8,97	0,33	9,34	100
Spectrum 2		73,92	5,1	6,65	9,01	0,24	5,07	100
Spectrum 3	19,49	57,57					22,94	100
Spectrum 4	8,12	66,41	0,83		0,22		24,43	100
Spectrum 5	19,84	57,45					22,71	100
Spectrum 6	13,46	68,92		0,08	0,25		17,29	100
Spectrum 7		77,89	3,51	4,1	6,98	0,28	7,24	100
Spectrum 8	6,81	64,6			0,14	0,32	28,14	100
Spectrum 9	11,62	56,79	4,61	0,08			26,9	100
Spectrum 10	8,76	60,06					31,18	100

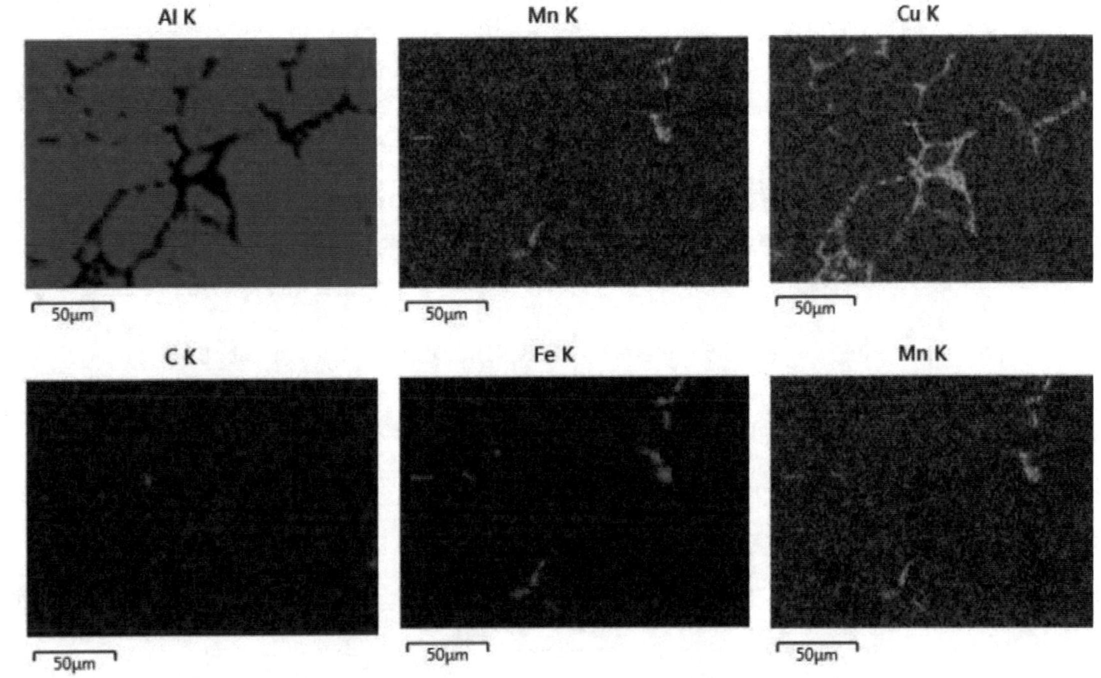

Figure 6. Chemical composition of grain-boundary phase separations in the D16 alloy sample after 18 hours of exposure in the furnace, EDAX

4. Conclusions

Fairly wide and versatile information about the aluminum alloy D16-duralumin was obtained. Alloy requires exposure during the technological process as shown by the experiment the most optimal time is 4 hours or 8 hours, depending on what goal is pursued. One of the important conditions is the heating temperature at which no oxidation or melting of the surface should occur. There should also be no deformations as a result of mechanical processing. The experiment was successful which led to excellent results. The developed process which shows all the details D16 describes some regularities of changes in the pattern of the interaction structure, the possibility of determining the chemical composition of the grain boundary phase precipitates in the sample and the experimental dependence of the change % of the area of the micro section occupied grain-boundary phase, from the time of exposure.

Acknowledgements

Thanks to Don State Technical University for the excellent organization of the conference.

References

[1] ASM Handbook 2004 Vol. 9: Metallography and microstructures, ASM International. 1688-1787
[2] Arzamasov B. N. Handbook of structural materials / Arzamasov B. N. Solovieva T. V. Moscow: Publishing house of Bauman Moscow state technical University, 2006. - 636 p., Ill. - ISBN 5-7038-2651-9.
[3] Warmuzek M. 2004. Vol 9, ASM Handbook, ASM International. P. 711-751 (Malgorzata, Warmuzek, Foundry Research Institute, Cracow, Poland).

IOP Conf. Series: Materials Science and Engineering **900** (2020) 012011 doi:10.1088/1757-899X/900/1/012011

Improving the efficiency of the path - rolling stock system based on the implementation of anisotropic frictional bonds

V V Shapovalov, V I Kolesnikov, P V Kharlamov, R A Kornienko*, A M Petrik

Rostov State Transport University, Rostov-on-Don, Russia

*e-mail: Kornienkoroman23@mail.ru

Abstract. Unlike the materials widely represented in domestic and foreign scientific, technical and reference literature, the article theoretically substantiates the need to take into account in laboratory and bench tests of mechanical systems with friction units the presence and interconnection of dynamic processes occurring in a frictional contact and in a mechanical quasilinear subsystem; ensure identical: parameters of macro- and micro-roughness of contacting surfaces, frequencies and forms of natural vibrations, physical and mechanical properties of tribocontacts, as well as dynamic (integral) characteristics of a natural object and its physical model. The article presents the methods of physical and mathematical modeling and tribospectral identification of friction processes. The features of these methods and an example of dynamic monitoring of a mobile mechanical system using the example of a friction system: "way - rolling stock" are consideredThe first section in your paper.

1. Introduction

Issues of increasing the volume and stability of locomotive traction force play a key role in improving the efficiency of the railway industry, which is also reflected in the above concept. In order to increase traction, a train with two locomotives located in front and behind the train is used; quartz sand is used as a friction activator. Possible misalignment of traction forces when using a locomotive-pusher leads to an increase in resistance to movement of the train, creates additional forces in the "wheel flange-rail" contact, increases the intensity of the transverse creep forces and the wear intensity of the wheels and rails [1-3]. The use of sand, which is an active abrasive, leads to a sharp increase in the wear rate of wheels and rails, as well as to losses of up to 5% of traction power associated with the energy required to destroy sand particles falling under the wheels. On prolonged slopes, due to the active supply of sand, ballast sanding occurs, which reduces the repair time of railway maintenance. the way. A single refueling of locomotive with sand provides a turnover shoulder of not more than 1.5 thousand km, with the necessary shoulder reaching up to 8 thousand km.

To solve these problems, a technology has been developed for controlling the frictional contact of the wheels of a locomotive. An on-board automatic control and management system (ACMS) with friction surface modifier (FSM) briquette feed devices installed on a locomotive predicts the onset of the slip level with the aim of preliminarily turning on the modification system and eliminating traction implementation modes at an increased slip speed level. FSM briquette feeders with ACMS systems are installed on all wheelsets of a locomotive, which implements autonomous (axial) control of power units [1-4].

2. Organizations of dynamic monitoring of processes wheel-rail interaction

The quality criteria of the friction-mechanical system are set up on the basis of frequency characteristics. Frequency characteristics do not consider the type of transient process, are based on some frequency properties of the system and are designed to evaluate parameters intended to limit certain parameters [2-6].

The following characteristics of the friction-mechanical system can be identified by the frequency characteristics [2,3,7-10]:

1) stationary value of the transmission coefficient

$$Wxy(0) = A(0) = f\text{тр} = \frac{F\text{тр}}{N} \approx \frac{C_x x}{C_y y} \quad (1)$$

where $W_{xy}(0)$ is the value of the complex transfer coefficient at a value of the oscillation frequency ω, equal to zero, which corresponds to the transfer coefficient in a stationary steady motion of the friction-mechanical system;

$A(0)$ is amplitude value on the graph of the amplitude-frequency curve;

$f_{mp.}$ is the value of the friction coefficient in a stationary steady motion;

$F_{mp.}$ is the friction force, N;

N is the normal load, N;

C_x is the coefficient of elasticity of frictional bonds in the direction of relative displacement, N/m;

C_y is coefficient of elasticity of friction bonds in the direction of relative approach of the contacting friction surfaces, N/m;

x is the amplitude of tangential displacement (deformation of active volumes) of the friction surface;

y is the amplitude of approach of the contacting friction surfaces.

2) the magnitude of the *stability margin in amplitude* [2, 10], characterizing the possibility of limiting the inertial effects that contribute to breaking of the friction coupling in tribocontact:

$$L = 20\lg\frac{1}{|W(i\omega_{c2})|}, \quad (2)$$

where ω_{c2} is cutoff frequency corresponding to the point of intersection of the hodograph of the amplitude phase function $W(i\omega_{c2})$ dynamic coefficient of friction of the real axis, at which the frequency transfer coefficient of the open system is equal to unity, and the phase shift is –180°;

$W(i\omega_{c2})$ is amplitude phase function of the dynamic coefficient of friction;

1 is the critical value of the amplitude at which the amplitude of the tangential displacement of the friction-mechanical system is equal to the amplitude of the load;

3) the value of the *phase stability margin* Ψ [2, 3, 10], allowing to limit the inertial and dissipative components of the frictional interaction forces that contribute to the convergence of the contacting surfaces of friction, increase contact stresses and temperature) - a value determined at a frequency at which the complex coefficient of friction is equal to unity, i.e. $|W(i\omega)| = 1$.

4) the frequency indicator of the oscillation M, characterizing the tendency of the friction-mechanical system to oscillate, and its value is limited in the range from 1.1 to 1.5

$$M = \frac{|\Phi(i\omega)|_{\max}}{|\Phi(0)|} = \frac{\left|\dfrac{W(i\omega)}{1+W(i\omega)}\right|_{\max}}{\left|\dfrac{W(0)}{1+W(0)}\right|} = \frac{A(\omega)_{\max}}{A(0)} = \frac{m^2+1}{2m}, \quad (3)$$

where $\Phi(i\omega)$ is amplitude phase frequency curve of a closed system using unit feedback via tribocontact

$$\Phi(i\omega)=\frac{W(i\omega)}{1+W(i\omega)};$$

$W(i\omega)$ is original open-loop frequency transfer function;

$A(\omega)_{max}$, $A(0)$ is accordingly, the maximum amplitude of the lowest frequency harmonic and the amplitude of the steady state;

m is root indicator of system stability margin.

5) cutoff frequency ω_{c3} at $A(\omega_{c3}) = A(0)$, determining the duration of the transition process. The higher the cutoff frequency, the shorter the transition process and the greater the speed

$$\omega_{c3} = \omega\Big|_{A(\omega)=A(0)},$$

where $A(\omega)$ is amplitude frequency function;

$A(0)$ is the value of the coefficient of friction in the steady (stationary) state.

6) transmission band frequencies ω_0 at $A(\omega_0) = 0{,}707 \cdot A(0)$, which evaluates the noise immunity of the friction-mechanical system. The transmission ban should not be too wide

$$\omega_0 = \omega\Big|_{A(\omega)=\frac{\sqrt{2}}{2}A(0)}.$$

7) the integral quality criterion I of the Heaviside function [2, 3, 10], which characterizes the dynamic error that occurs when the system responds to the input master action of the type of the Heaviside function 1 (t). The estimate is calculated in the frequency domain based on the Parseval theorem and the Rayleigh formula. Based on these methods, the integration of the square of the time function ranging from zero to infinity is replaced by the integration of the square of the module of the complex amplitude of the closed system, and the Fourier image of the deviation under study determines the error of the transient dynamic process ε

$$I = \int_0^{\infty}(h_{st} - h(t))^2 dt \equiv \frac{1}{\pi}\int_0^{\infty}\frac{|\Phi(0)-\Phi(i\omega)|^2}{\omega^2}d\omega$$

where h_{st} is steady-state (stationary) value of the Heaviside transition function (with time $t = \infty$).

8) integral quality criterion I' of Dirac function [2, 3, 10], which characterizes the dynamic error that occurs when the system responds to the input defining action of the type of the Dirac function δ(t), that is, an impulse effect. The estimate is also calculated based on the Parseval theorem and Rayleigh formula:

$$I' = \int_0^{\infty}|w(t)|^2 dt \equiv \frac{1}{\pi}\int_0^{\infty}|\Phi(i\omega)|^2 d\omega, \quad (4)$$

where $w(t)$ is values of Dirac impulse function at time t.

9) the integral value of the elastic-inertial components I_C, contributing to an increase in potential energy. The lower value I_C, the less inertia of the system has on the dynamic approach of the contacting surfaces of friction to each other, an increase in contact stresses and surface temperature, the development of plastic deformations, and also on the stability of the dynamics of the processes of friction [2, 3, 10].

$$I_C = \int_0^{\infty} P(\omega)d\omega, \ if\ P(\omega) \geq 0,$$

where $P(\omega)$ is real frequency function.

10) the integral value of the inertial components I_m [2, 3, 10], contributing to an increase in kinetic energy, loss of stability of the frictional bond, decrease in system stability. The less the value of I_m, the lower the inertia of the system and the higher the stability of the friction processes

$$I_m = \int_0^\infty |P(\omega)| d\omega \text{, } if\ P(\omega) < 0.$$

11) the integral value of resistance to vibrations I_{conp} [2, 3, 10], characterizing the dissipative properties of the system. The larger the integral estimate, the higher the amplitude of the oscillations caused by the combination of inertial and dissipative components of the frictional interaction forces. The lower I_{conp} value, the less kinetic energy is spent on friction processes in the frictional bonds of the friction-mechanical system, the lower the temperature fluctuations

$$I_{conp} = \int_0^\infty |Q(\omega)| d\omega \text{, } if\ Q(\omega) \leq 0,$$

Where $Q(\omega)$ is imaginary frequency function.

12) the integral magnitude of the forces caused by frictional self-oscillations [2, 3, 10], $I_{фр.к.}$ characterizes the dissipative properties of the system when the vector of the resistance to motion is co-directional with the velocity vector. Its increase contributes to a decrease in stability and stability, breaking friction bonds. A significant increase in this estimate may indicate a significant increase in the relative sliding rates of friction surfaces while overcoming local or local resistances caused by the transition from elastic to elastoplastic deformations or from elastoplastic to plastic deformation of contacting surfaces, wear, and thermal damage to the friction-mechanical system. The lower the $I_{фр.к.}$value, the more stable the friction-mechanical system

$$I_{фр.к.} = \int_0^\infty Q(\omega) d\omega \text{, } if\ Q(\omega) > 0.$$

13) Integral damping coefficient estimation I_ξ, characterizing the ratio of the dissipative and conservative energy of the friction-mechanical system and approximately corresponds to its linear value ξ

$$I_\xi \approx \frac{1}{\sqrt{1+\frac{1}{(\frac{I_{comp}+I_{frk}}{I_c+I_m})}}}; \quad \xi = \frac{n}{\omega_0} = \frac{\beta}{\sqrt{4mC}} = \frac{\beta}{\beta_{кр}},$$

Where I_{conp} is the integral assessment of resistance to vibrations;

$I_{frк}$ is the integral assessment of forces caused by frictional self-oscillations;

I_C is the integral assessment of elastic-inertial components;

I_m is the integral assessment of inertial components;

n is the vibration damping coefficient, c^{-1};

ω_0 is the natural frequency, c^{-1};

β is the equivalent vibration resistance coefficient, N·s/m;

m is the reduced mass of contacting surfaces involved in friction, kg;

C is the reduced coefficient of elasticity of contacting surfaces, N/m;

β_0 is the equivalent critical value of the coefficient of resistance to vibrations, at which the oscillatory character is replaced by a monotonically damping (aperiodic).

Having determined the frequency characteristics of the friction-mechanical system, according to the real function and integral V.V. Solodovnikov the transition function is calculated that characterizes the free component of the vibrational states of the system.

3. Conclusion
The temporal characteristics in the form of a transition function make it possible to evaluate the nature of the change in the coefficient of friction in the friction-mechanical system from the impact on the object of study (mathematical model) of the standard Heaviside function 1 (t). If the system meets the specified quality requirements, then it will satisfy these requirements under any other arbitrary impact.

The publication was carried out as part of a grant from RFBR № 18-58-00026.

References
[1] Shapovalov V.V. Actual Problems of Modern Tribotechnology and Ways to Solve Them Higher Education News (ser. Mashinostroenie). 2015. № 1. p. 64-75
[2] Patent of RF No. 2343450. Method for testing friction units / V.V. Shapovalov, I.A. Mayba, A.L. Ozyabkin, P.V. Kharlamov [et. al.] ; BI No. 1. –Published on 10.01.2009.
[3] Patent of RF No. 2517946, MPK G 01 N 3/56 (2006.01). Method for dynamic monitoring of frictional mobile systems / V.V. Shapovalov, A.L. Ozyabkin, P.V. Kharlamov [et. al.]. No. 2517946; application of 05.04.2012; published on 10.06.2014, Bul. No. 16; priority 05.04.2012, № 2012113329/28.
[4] Ozyabkin A.L. Monitoring the hydrothermodynamics of the frictional contact of the mobile tribosystem / A.L. Ozyabkin, I.V. Kolesnikov, P.V. Kharlamov / Friction and lubrication in machines and mechanisms. 2012. №3. p. 25-36.
[5] Loktev A.A. Simulation of the railway under dynamic loading. Part 1. Ray method for dynamic problem/A.A. Loktev, E.A. Gridasova, E.V. Zapolnova//Contemporary Engineering Sciences. -2015. Vol. 8. № 18. P. 799-807.
[6] Evans J. & Berg, M. Challenges in simulation of rail vehicle dynamics, Vehicle System Dynamics. 2009. Vol. 47. P. 1023-1048.
[7] Dumitriu M. Modeling of railway vehicles for virtual homologation from dynamic behavior perspective. Applied Mechanics and Materials. 2013. Vol. 371. P. 647-651.
[8] Yang Y. C., Chu S. S., Chang W. J., and Wu T. S. Estimation of Heat Flux and Temperature Distributions in a Composite Strip and Homogeneous Foundation//Int. Commun. Heat Mass Tran. 2010 (37), no. 5, 495-500.
[9] Wolter K.U. & Zacher, M. & Slovak, B. Correlation between track geometry quality and vehicle reactions in the virtual rolling stock homologation process. In: 9th World Congress on Railway Research, May 22-26. 2011.
[10] Amplitude-phase-frequency analysis of critical states of friction systems: Monograph / V.V. Shapovalov, A.V. Chelokhyan, I.V. Kolesnikov, A.L. Ozyabkin, P.V. Kharlamov. 2010. 383 p.

Energy efficiency assessment of anthropomorphic manipulator control algorithm

K I Boldareva[1],*, A D Lukyanov[1], E H Abderrazzak[2]

[1]Don State Technical University, Laboratory of local automation and embedded systems, Rostov-on-Don, Russia,
[2]Ministère Équipements Transport Logistique Eau, MEQTLE, Rabat, Maroc

*e-mail: gorianina.k@yandex.ru

Abstract. Currently, developers of robotic systems are very interested in intensively developing unmanned technologies, the concept of which is to use robots to perform routine, harmful and dangerous types of work without direct human participation, which is the key to ensuring safety and high efficiency of solving tasks. Active research is conducted in the field of artificial intelligence and the development of anthropomorphic robots (AR). Modern ARS are not yet able to completely replace a person when performing complex tasks in a dynamic environment. The inability of such robots to completely replace a human is due to the short duration of work in offline mode. The reasons for the short duration of the AP are the low capacity of modern batteries and high power consumption of software and hardware components of the robot. The solution to this problem is possible by developing and modifying mathematical methods and algorithms for performing target operations in order to minimize energy consumption. The article assesses the energy efficiency of the algorithm for copying control of an anthropomorphic manipulator.

1. Introduction

Modern Autonomous anthropomorphic robots face a number of problems, one of which is the low duration of their work. This type of robot is designed to operate in conditions that are dangerous to human life – in space, under water or under high radiation, to perform irregular operations – rescue operations, maintenance of space equipment, equipment for nuclear facilities, research missions, etc. Execution of the main target operations of these robots is performed by Executive devices-anthropomorphic manipulators (AM).

The power consumption of manipulators depends on the way of its movement, so the application methods energy-efficient path planning will increase the duration of the work and time of useful work (less need to spend time recharging or changing the energy element) and the battery life (a limited number of cycles charge/discharge).

In modern robotics, the problem of limited energy resources of autonomous anthropomorphic robots (AR) is relevant. Numerous scientific researches are devoted to the solution of the presented problem, among the most known tendencies it is possible to allocate the following: studying of new strong and easy materials, creation of perspective powerful power sources, variations of odification and optimization of AR structure, the solution methods development of energy-efficient movement management problems.

IOP Conf. Series: Materials Science and Engineering **900** (2020) 012012 doi:10.1088/1757-899X/900/1/012012

The authors are particularly interested in the problem of planning an energy-efficient motion path (MP) for redundant AR manipulators. In this area, a number of papers dedicated to the development of methods and algorithms for planning the energy-efficient path of the AR manipulator in the work area with an obstacle [1-4].

The tasks of motion path planning of manipulators with redundant mobility in the working area with obstacles have been developed since the 1980s [5]. Motion generation for redundant manipulators is usually considered with using kinematic control [6]. The task of finding redundant manipulators motion path is often to find any valid way from an infinite number of options. The solution of this problem is to construct a polyline curve of the displacement line for its subsequent approximation by cubic splines and the transition to the solution of dynamics problems. The tasks of optimal motion path constructing according to the given selection criteria are more relevant in applied research. The basis of problems of optimum values search is put in [7-8]. When solving the optimal path-finding problem according to the established quality criteria, a problem of high computational complexity arises [9-10].

There are many efficient but computationally complex methods for planning the path of a manipulator. Two approaches to trajectory planning are widely covered in the literature - the approach based on graph theory, as well as spline interpolation. The first approach is associated with great computational complexity, and the second – with the complexity of selecting reference points, excessive movement for a small number of them, and the complexity of calculations for a large one.

Planning the trajectory of movement of redundant manipulators based on evolutionary approaches is considered in [11-13]. The proposed solutions are aimed at obtaining an accurate solution both in offline mode and in real time, but do not take into account the power consumption of the manipulator motors when moving.

Methods for constructing the trajectory of the manipulator movement based on learning the natural movements of the human hand are presented in [14-15]. At the same time, there are problems with the computational complexity of the proposed solutions, and the natural optimization of human hand movement does not meet the requirements for performing the operation with minimal energy consumption.

Numerical methods for solving the trajectory planning problem are presented in [16-17]. The execution of trajectory generation procedures in these works has a high computational complexity, which requires high performance of the computing system.

Based on this, it is an urgent task and purpose to develop energy-efficient path planning methods for manipulators operating in real time or close to it.

2. Analyze of problems of energy consumption of manipulators of autonomous anthropomorphic robots when performing target operations

The current pace of development of anthropomorphic robotic systems is due to the tendency to replace a person when performing potentially dangerous work. Thus, researchers face a number of tasks that will allow them to replace a person in various fields of activity, including rescue operations, space missions, work in radiation conditions, during fires, military operations, and natural disasters. One of these tasks is to increase the duration of target operations performed by the robot in offline mode.

The structure of the AR's power consumption depends on the number of operations performed in its modules [18]. The modular structure of the AR implies the presence of transport (pedipulators), Executive (manipulators), service (sensors, sensors, displays, etc.) and control (processors, controllers, etc.) modules.

Existing methods of reducing the power consumption of AR manipulators can be divided into two types: design modification and modification of motion control algorithms. Let's consider the advantages and disadvantages of the presented methods. The advantage of reducing energy consumption by modifying the design of the AP manipulators and modifying the motion control algorithms is shown in table 1.

Modification of the AR design should be carried out at the design stage, since the replacement of components, parts and engines in existing samples is associated with great, sometimes insurmountable difficulties and requires recalculation of the system dynamics of the manipulator, compatibility, etc. Modification of motion control algorithms includes modification of methods for planning the path of movement and modification of methods for planning the trajectory of the manipulators of the anthropomorphic robot [19]. By the method of path planning of the manipulator refers to the planning of the route of the manipulator from the starting point *A* to the end point *B*, including intermediate points [20]. In other words, searches basic start and end coordinates of the united successive lines of the route and the values of generalized coordinates of the manipulator in the configuration space.

Table 1. Advantages and disadvantages of methods for reducing power consumption of the manipulator.

Method description	Advantages	Disadvantages
Facilitating the construction of the manipulator, i.e. using strong and light materials	Reducing the weight of the manipulator; reducing the requirements for the power of electric motors	Significant increase in the cost of construction
Choice of economical electric motors with high efficiency	Reduced engine power consumption	Increased engine wear and construction costs
Reducing the number of degrees of mobility of the manipulator	Simplified design, reduced cost	Reduced mobility of the manipulator, possible increase in the number of manipulator operations
Modification of methods for planning the path of movement of the manipulator	Reduced energy consumption due to optimal travel path	There may be a slight increase in the path of movement of the manipulator
Modification of methods for planning the trajectory of manipulators	Reducing energy consumption by building an optimal path for the movement of links	It is possible to reduce the speed and accuracy of performing target operations with the manipulator

Planning the path of movement of the manipulator includes continuous intervals between points, the movement between which is carried out using methods of planning the trajectory of movement. To solve these problems, it is possible to use a combination of solving the forward and reverse dynamics problem in conjunction with forecasting methods [21].

The use of these methods is also relevant for reducing the power consumption of manipulators [22].

Among the disadvantages of these methods, we can highlight a decrease in indicators for evaluating motion control algorithms, which include accuracy, path length and speed, and computational complexity.

The main operations to reduce the energy consumption of manipulators in motion control algorithms are solved at the stage of planning the trajectory of the manipulator, which is not correct, since the problem of energy efficiency is also relevant for methods of planning the path of movement. The search for the shortest path is not optimal due to the possibility of performing more movement by the link located at the base of the manipulator, the engine of which is more powerful and consumes more energy.

3. Problems of copying control of an anthropomorphic manipulator

One of the main requirements for the copying control system is that the manipulator reproduces the rotation angles of the joints of the master device with sufficient accuracy and minimal time delay.

When using digital systems, there is inevitably a loss of accuracy due to digital-to-analog conversion of control signals (signal sampling by time and quantization by level).

Currently existing angle sensors for copying devices (encoders) have a bit rate of 12 or more bits. The use of encoders of this size allows you to determine the angle of rotation with an accuracy of at least 0.05 degrees with acceptable values of the angle of rotation in the range from 0 to 180 degrees. Such accuracy indicators are sufficient to solve the problems of copying control of an anthropomorphic manipulator, so the error of encoders in this study can be ignored.

Recovery of an analog signal from a sequence of discrete samples can be performed with high accuracy using Kotelnikov's theorem. However, the theorem can be applied in the field of copy control is limited by the fact that the recovery of the value of the signal level in the current time you must have information not only about the preceding and the subsequent counts, which contradicts the requirement for low latency and principles of real-time systems in General.

Figure 1 shows the kinematic scheme of the anthropomorphic manipulator considered in this article. The manipulator has 7 degrees of rotational mobility of the A1 – A7. The designations B1, B2 B3 correspond to the shoulder, elbow and wrist nodes, B4-the working end.

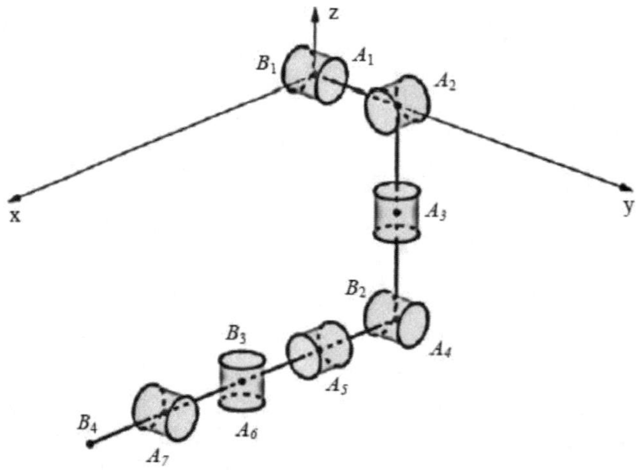

Figure 1. Kinematic diagram of an anthropomorphic robot manipulator with 7 degrees of mobility

The master device, which is used to copy the control of an anthropomorphic manipulator, is an exoskeleton of the upper extremities, which has a kinematic scheme similar to the kinematic scheme of the manipulator. An example of such devices is the master device of a copying manipulator [23].

The system may control an anthropomorphic manipulator, except setting device and of the manipulator also includes a computer control module anthropomorphic manipulator device video surveillance (video), the display device of the video (monitor) and a communication channel through which the command transmission control module and of the video generated by the video monitoring device, on a computer operator.

Each command for the control module of an anthropomorphic manipulator is a vector θ of the rotation angles of the joints corresponding to the seven degrees of mobility of the manipulator:

$$\theta = \{\theta_1, \theta_2, \theta_3, \theta_4, \theta_5, \theta_6, \theta_7\} \quad (1)$$

If the master and manipulator use the same rotation angle sensors (encoders), the angles θ_i, $i = 1..7$ can be represented directly in the code used by the encoders to reduce the latency and computational costs associated with converting data to an intermediate format.

Based on the received commands, the control module generates signals for the electric motors, tracking the angle of rotation of each joint by means of encoders connected to the electric motors.

The total delay in the am copying control system is caused by delays that occur at the following stages of the system operation:

– receiving and processing data about the operator's hand position;

- forming commands for the AM control module;

- formation of data packets corresponding to the information exchange Protocol used;

– sending commands over the communication channel;

– receiving and processing commands by the AM control module, generating control signals for electric motors of Executive devices;

- changing the spatial position of the components of the mechanical AM system according to the received commands;

– image capture and encoding of the video stream by the video monitoring device;

– transmitting a video stream over a communication channel;

– receiving and decoding the video stream;

– output the image to the display device (monitor).

The most specific of these delays for copying control of an anthropomorphic manipulator are:

– receiving and processing data about the operator's hand position;

- forming commands for the AM control module.

The absence of mechanisms for reducing the level of sampling error in the copying control system can lead to undesirable vibrations associated with a sharp transition of the spatial configuration of the manipulator from the state corresponding to the (k-1)-th sample of discrete time to the current state corresponding to the k-th sample. Increasing the sample rate minimizes this effect, but also increases the bandwidth requirements of the communication channel.

4. Conclusion

The use of analytical methods of path planning is relevant in problems of path planning at a large sampling step, but this negatively affects its computational complexity. Thus, it is important to develop this direction in the direction of reducing the complexity of the algorithm, for example, when presenting an obstacle with a geometric primitive and developing energy efficiency criteria when searching for a route.

To search for a path, the criteria for finding the shortest route or any acceptable solution is used, but the problem of energy consumption is not solved. Therefore, since the shortest path is not always optimal in terms of energy consumption, the task of developing an energy-efficient method for planning the travel path for an anthropomorphic robot manipulator in a work zone with a stationary obstacle in real time is relevant.

Abstract similarity of the operator's hand positions and the AR can be implemented in various ways. Similarity can be achieved due to the same rotation angles in the operator's hand joints and AR joints.

References

[1] Petrenko V I 2018 Method of planning the point motion trajectory in space with an obstacle based on the iterative piecewise-linear approximation *System control, communication and security* № 1 pp 168-182

[2] Petrenko V I 2018 Mathematical model of finding optimal Euler angles for motors three-link manipulator *Modern science: Actual problems of theory and practice. Natural and technical Sciences* № 3 (1) pp 67-74

[3] Petrenko I V 2018 Method of planning the manipulator optimal trajectory in a three-tier volumetric space with an obstacle *Bulletin of the Dagestan state technical University. Technical science* vol 45 №1 pp 98-112

[4] Antonov V O 2018 Analysis of the computational complexity of the iterative piecewise linear eneration method of the three-link anthropomorphic manipulator trajectory in the space with an obstacle *Proceedings of the Southwest state University* vol 22 №3 (78) pp 13-29
[5] Maciejewski A A and Klein C A 1985 Obstacle avoidance for kinematically redundant manipulators in dynamically varying environments *Intl. J. Robotics Research* vol 4 pp 109-117
[6] Siciliano B 1990 Kinematic control of redundant robot manipulators: A tutorial *J. Intell. Robot. Syst* vol 3 pp 201–212
[7] Baillieul J 1985 Kinematic programming alternatives for redundant manipulators *Proc. IEEE Int. Conf. Robot. Autom* (St. Louis, MO, USA) pp 722–728
[8] Chang P H 1987 A closed-form solution for inverse kinematics of robot manipulators with redundancy *IEEE Trans. Robot. Autom* vol RA-3 pp 393–403
[9] Psychopof V H and Medvedev M Y 2011 Control of mobile objects in certain and uncertain environments *Moscow: Science* p 350
[10] Yusupova N I, Shakhmametova G R and Kamilyanov A R 2007 Search for trajectories of the multi-link manipulator in a complex three-dimensional space *Bulletin of the Ufa state aviation technical University* №9 (2) pp 71-75
[11] Lin H I 2014 A fast and unified method to find a minimum-jerk robot joint trajectory using particle swarm optimization *Journal of Intelligent and Robotic Systems: Theory and Applications* vol 75 pp 379-392
[12] Qi R, Zhou W and Wang T 2014 An obsta-cle avoidance trajectory planning scheme for space manipulators based on genetic algo-rithm *Jiqiren/Robot* vol 36 pp 263-270
[13] Xidias E K 2018 Time-optimal trajectory planning for hyper-redundant manipulators in 3D workspaces *Robotics and computer-integrated manufacturing* №50 pp 286-298
[14] Abu-Dakka F J, Valero F J, Suner J L and Mata V 2015 A direct approach to solving trajectory planning problems using genetic algorithms with dynamics in complex envi-ronments *Robotica* vol. 33 pp 669-683
[15] Ren Z W, Zhu Q G and Xiong R 2015 Trajectory planning of 7-DOF humanoid manipulator under rapid and continuous reaction and obstacle avoidance environment *Acta Automatica Sinica* vol 41 №6 pp 1131-1144
[16] Howard T, Pivtoraiko M, Knepper R A and Kelly A 2014 Model-predictive motion planning *IEEE Robotics and Automation Magazine* vol 21 №1 pp 64-73
[17] Chen Y J, Ju M Y and Hwang K S 2017 A virtual torque-based approach to kinematic control of redundant manipulators *IEEE Transactions on Industrial Electronics* vol 64 № 2 pp 1728-1736
[18] Makarov A B 1999 Analysis and optimization of modular structures of technological robots with double hinges p 184
[19] Abueva F B, Pyzhevskii D E and Antonov V A 2016 Control of mobile manipulation robots (Stavropol: NCFU Publishing house) pp 559-561
[20] Madhevan B and Sreekumar M 2018 Identification of probabilistic approaches and map-based navigation in motion planning for mobile robots *Academy Proceedings in Engineering Sciences* №43(8) pp 1-18
[21] Petrenko V I and Antonov V O 2018 Algorithm for solving kinematics and dynamics problems when controlling an anthropomorphic manipulator via a copying-type control device (Stavropol: NCFU Publishing house) pp 11-16
[22] Mikhailov E P, Woodwood A N, Konyukh A B and Onufrienko V S 2014 Positional control of mobile robot manipulators *Lifting and transport equipment* №1(41) pp 82-89
[23] Batrashkin A P, Bogdanov A A, Iksanov M R, Kutlubaev I M and Permyakov A F Setting device of the copying manipulator RU 169 864 U1

Find the optimal trajectory of the secondary discrete part to ensure efficient processing of cellulose fiber

D N Kuimov [1,*], M S Minkin [2]

[1]Don State Technical University, Department of scientific research, 344000 Rostov-on-Don, Russia
[2]Don State Technical University, Department of automation of production processes, 344000 Rostov-on-Don, Russia

*e-mail: kuimov_d@list.ru

Abstract. This work presents the results of investigation of trajectories of elements implemented by electromechanical converter with discrete secondary part. The efficiency of implementation of various trajectories was evaluated.

1. Introduction

Modern polymeric materials, including cellulose and cellulose-containing ones, are complex heterogeneous (multicomponent and multiphase) systems whose properties are determined by the chemical structure of the components, the nature and intensity of interaction between them, as well as the micro - and macrostructure of the material. One of the promising ways to control these characteristics and, as a result, change the properties of cellulose-containing composite materials is to use nanoscale elements of the cellulose structure-nanocrystalline and nanofiber cellulose as a modifying component.

However, in order to fully utilize the cellulose fiber, it is necessary to weaken the complex structure of the cellulose by pretreatment. To date, a large number of different methods of pretreatment of cellulose fiber have already been developed. For pulp milling in industrial conditions, various types of grinding devices are most often used. Mills are widely used as a grinding device, which is confirmed in the published results of scientific works of recent years [1,2]. The mass is ground between the walls of the housing by a rotating disk with knives that have a smooth surface. There are studies in which ultrasound or conventional heat treatment [3,4], and sometimes a combination of different methods, are used to destroy the structure of the cellulose fiber, which undoubtedly increases the cost and complicates the processing process.

Among the methods presented above, hydrodynamic cavitation as a new promising method is the least studied and studied.

2. Main part

The efficiency of hydrodynamic cavitation is due to local increases in pressure and temperature. At the same time, the high energy effect depends on the formation, growth and collapse of cavitation cavities (bubbles) generated by pressure drops in the liquid raw material. In a number of publications for generating hydrodynamic cavitation, various configurations of diaphragms with holes with a diameter of 1 mm appear [5]. The technology is based on the passage of liquid processed raw materials through holes with the occurrence of cavitation cavities at their boundaries.

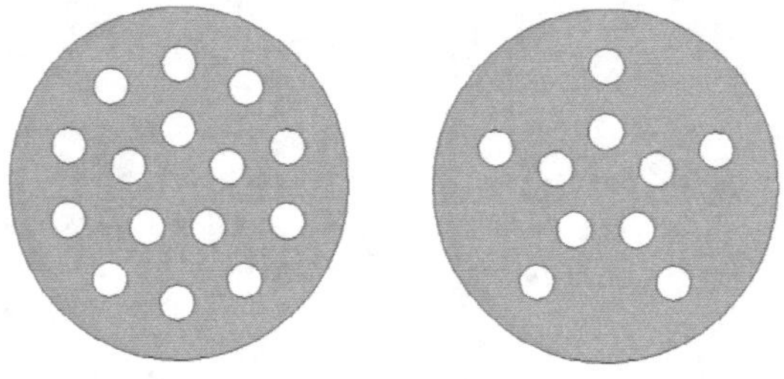

Figure 1. Schematic representation of diaphragms

In this study, an electromechanical converter with a discrete secondary part (ECDSP) was used to generate hydrodynamic cavitation in the pre-treatment of cellulose fiber. Unlike diaphragms, ECDSP is able to create a cavitation cloud under certain conditions throughout the entire volume of the working chamber, significantly accelerating the processing process, while increasing the energy efficiency of the process.

In addition to hydrodynamic cavitation, the treatment in ECDSP is accompanied by a radome of additional energy effects, including the impact of elements, magnetostriction and friction [6, 7].

Figure 2. Section of electromechanical converter with discrete secondary part

In ECDSP, the secondary part represented by a large set of ferromagnetic elements moves in the treated medium under the influence of an external electromagnetic field. In turn, electromagnetic field, its force and direction is determined by combination of connected concentrated coils arranged in inductor slots.

The movement of the elements at a speed higher than critical causes a vacuum of the liquid. In the vacuum phase, a gap is formed in the liquid in the form of a cavity, which is filled with saturated liquid vapor. In the compression phase, under the influence of increased pressure and surface tension forces, the cavity collapses. A significant effect of cavitation processes is associated with a high concentration of energy released during the collapse process in the treated medium. At the moment of collapse, the gas pressure and temperature reach significant values, and according to some data reach 100 MPa and 1000 °C, respectively [8].

Thus, the first most important parameter of pulp grinding technology in ECDSP is the achievement of maximum speed by moving elements. A feature of the present device for implementing pulp pre-grinding technology is the possibility, when including various combinations of inclusion of coils or coil groups, of implementing unique motion paths inaccessible to similar installations, referred to in most scientific literature as vortex layer activators. In accordance with the classification proposed in [9] for a magnetic system of inductor machines similar to the configuration of a magnetic system of an electromechanical converter with a discrete secondary part, the efficiency of the device depends on the selected phase switching algorithm. Each individual path of the ferromagnetic elements may be useful

IOP Conf. Series: Materials Science and Engineering **900** (2020) 012013 doi:10.1088/1757-899X/900/1/012013

in the treatment of a particular liquid material. However, within the framework of this study, the speed of motion is of paramount importance, since achieving a speed close to or higher than a critical value allows intensifying hydrodynamic cavitation.

The first stage of the study was the modeling of various trajectories of the ferromagnetic element with parameterization of the speed and free run path. According to the previously presented works [10], the following system of differential equations was used to study the trajectory and motion parameters of the ferromagnetic element:

$$\ddot{x} = \frac{F_{ЭМx}}{m_Э^{'}} - \frac{1}{2m_Э^{'}} cS_М \rho_Ж \dot{x}\sqrt{\dot{x}^2 + \dot{y}^2}\,; \tag{1}$$

$$\ddot{y} = \frac{F_{ЭМy}}{m_Э^{'}} - \frac{1}{2m_Э^{'}} cS_М \rho_Ж \dot{y}\sqrt{\dot{x}^2 + \dot{y}^2} - \frac{gV_Э}{m_Э^{'}}(\rho_Э - \rho_Ж)\,. \tag{2}$$

The study of the electromechanical converter with a discrete secondary part was carried out for the four most optimal and implemented control algorithms in two stages, using instantaneous values of variables. The optimality of the implemented algorithms is due to restrictions on the heating resistance of the coils, since with an increase in the current density in the coils, it is possible to achieve the speed of ferromagnetic elements significantly higher in a short-term mode. In the four combinations presented, the degree of heating of the coils allows the plant to be used in a long-term mode without disrupting the operation and breakdown of the insulation.

The order of the study was divided into two stages. At the first stage, based on the results of calculations, the position of the working ferromagnetic element was fixed at the boundaries of the given areas and the speed and length of the free run of the ferromagnetic element to contact the working chamber was calculated. At the second stage, coordinates of each new position of the ferromagnetic working element in the space of the working chamber were recorded, and frame-by-frame movement of the working element was built.

Simulation results based on the frame-by-frame trajectory of the single ferromagnetic element are shown in fig. 3-6.

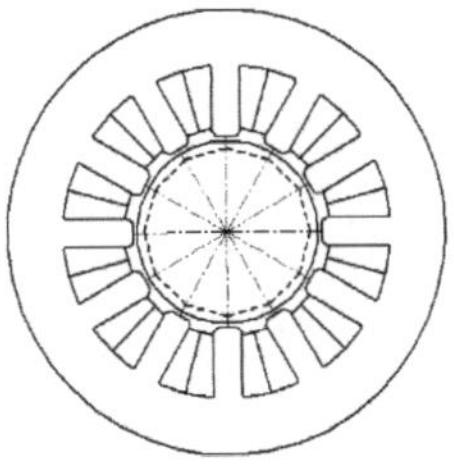

Figure 3. Model of motion path of unit ferromagnetic element A

Table 1. Energy characteristics of motion path A

I, A	10	11	12	13	14	15	16	17	18	19	20
V_m, m/s	2,6	2,9	3,7	4,7	5,2	6,1	7,1	7,5	8,2	8,7	9
V_a, m/s	0,6	0,7	0,76	0,91	1,1	1,2	1,5	1,7	1,9	2,1	2,4

IOP Conf. Series: Materials Science and Engineering **900** (2020) 012013 doi:10.1088/1757-899X/900/1/012013

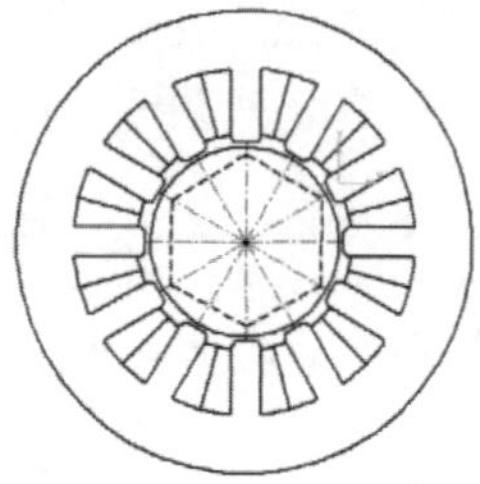

Figure 4. Model of motion path of unit ferromagnetic element B

Table 2. Energy characteristics of motion path B

I, A	10	11	12	13	14	15	16	17	18	19	20
V_m, m/s	3,1	3,4	4,1	5,0	5,8	6,6	7,4	7,9	8,7	9,3	9,8
V_a, m/s	1,5	1,9	2,2	2,6	2,9	3,2	3,5	3,9	4,1	4,4	4,6

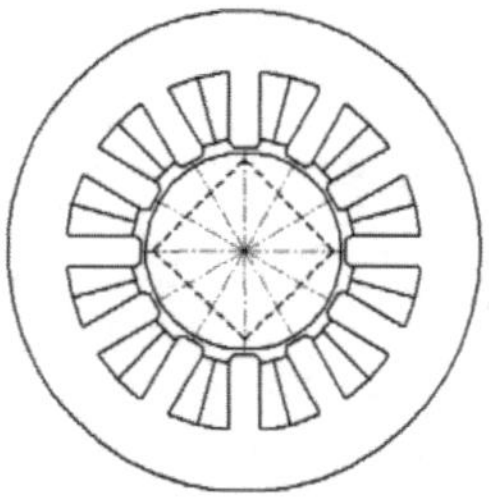

Figure 5. Model of motion path of unit ferromagnetic element C

Table 3. Energy characteristics of motion path C

I, A	10	11	12	13	14	15	16	17	18	19	20
V_m, m/s	3,2	3,5	4,2	5,3	5,9	6,8	7,8	8,3	9,2	9,7	10,2
V_a, m/s	0,5	0,6	0,7	0,89	0,98	1,1	1,3	1,6	1,7	1,9	2,1

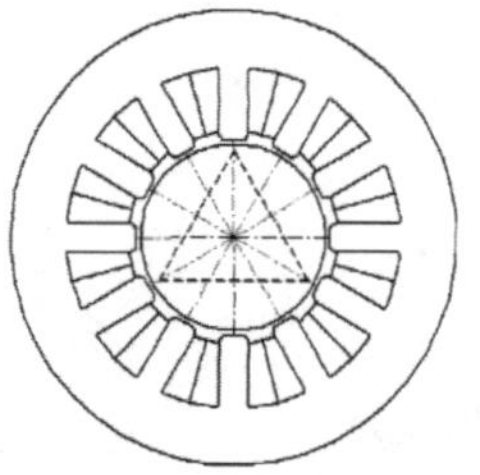

Figure 6. Model of motion path of unit ferromagnetic element D

Table 4. Energy characteristics of motion path D

I, A	10	11	12	13	14	15	16	17	18	19	20
V_m, m/s	3,5	3,8	4,4	5,6	6,2	7,2	8,1	8,7	9,5	10,1	10,3
V_a, m/s	0,6	0,64	0,71	0,82	0,94	0,97	1,1	1,4	1,5	1,7	1,9

According to a number of parameters of the studied trajectories, it can be concluded that in some cases an increase in the maximum speed does not lead to an increase in the length of the free run of the element with the speed necessary for the occurrence of a cavitation cloud in the path behind the element. Thus, in order to select the optimal path of motion, it is necessary to determine the fraction of the path in which a stable cavitation cloud will form behind the ferromagnetic element. The analysis of the movement of the single ferromagnetic element showed that the elements, when moving from one active pole to another, go through the acceleration and braking stages (fig. 7). The braking and acceleration path largely depends on the combination of coils included at each time point when the ferromagnetic element moves from one point to another. In this study, the proportion of the path of the element at which cavitation occurs was expressed through the trajectory coefficient.

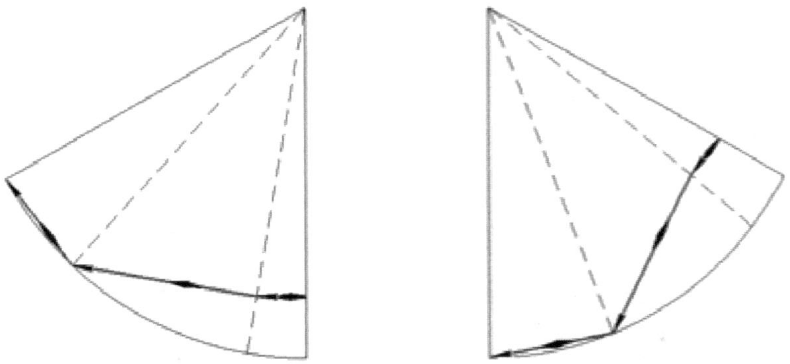

Figure 7. Motion of the element in ascending and descending trajectory

Trajectory analysis yielded some averaged data showing the proportion of path travelled by the ferromagnetic element at a velocity below the critical, expressed trajectory coefficient shown in table 5.

Table 5. Dynamics of ferromagnetic element motion

Trajectory type	Proportion of path travelled with cavitation
A	0,86
B	0,91
C	0,72
D	0,6

As a result, trajectory analysis yielded some averaged data showing the proportion of path travelled by the ferromagnetic element at a velocity below the critical, expressed trajectory coefficient shown.

3. Conclusion

The analysis of the possible trajectories of the ferromagnetic elements of the ECDSP made it possible to draw some conclusions regarding the implementation of the pulp treatment process by hydrodynamic cavitation. Despite the importance of the speed of motion in the implementation of cavitation, in the treatment process itself, the time of action of hydrodynamic cavitation on cellulosic fiber is no less important. In other words, the greater the proportion of path travelled by the ferromagnetic element with the cavitation cloud, the longer the hydrodynamic cavitation can affect the cellulosic fiber in one cycle of motion of the secondary discrete part. Increasing the fraction of the cavitation trajectory will reduce the total processing time of the cellulosic fiber. In this study, trajectory B was selected as the most optimal.

The reported study was funded by RFBR according to the research project № 18-29-18064.

References

[1] Gavrilova A.S., Manayenkov O.V., Filatova A. E., Investigation of the effect of ultrasound on microcrystalline cellulose, Bulletin of Tver state technical university, 2015, pp. 60-66.

[2] Sarymsakov A. A., Baltayeva M. M., Nabiev D. S., Rashidova S. sh., Yugay S. M. Dispersed microcrystalline cellulose and hydrogels based on it // Chemistry of plant raw materials. 2004, №2, pp. 11-16.

[3] Gharehkhani S. Basic effects of pulp refining on fiber properties – A review / S. Gharehkhani, E. Sadeghinezhad, S.N. Kazi, H. Yarmand, A. Badarudin, M.R. Safaei, M.N.M. Zubir // Carbohydrate Polimers. 2015. 115. P. 785-803.

[4] Kerekes, R.J. Characterizing refining action in PFI mills / R.J. Kerekes, R.M. Soszynski, P.A.T. Doo // Tappi Journal. 4(3). 2005. P. 9-13.

[5] Ruly Terán Hilares, Gabriela Faria de Almeida, Muhammad Ajaz Ahmed, Felipe A.F. Antunes, Silvio Silvério da Silva, Jong-In Han, Júlio César dos Santos, Hydrodynamic cavitation as an efficient pretreatment method for lignocellulosic biomass: A parametric study, Bioresource Technology 235 (2017), pp. 301–308.

[6] Kuimov D., Minkin, M. The electromechanical converter in the systems of desulfurisation of crude oil, *MATEC Web of Conferences*, 2017.

[7] Kuimov D., Pavlenko, A., Minkin, M., Optimization of parameters of the electromechanical converter with a discrete secondary part, *MATEC Web of Conferences*, 2017.

[8] Piersol I. Cavitation. Moscow: Mir, 1975. - 95 p.

[9] Shevkunova A.V. Choice of optimization method for the tooth zone of the valve-inductor motor, New science: Problems and prospects, 2016, N. 3-2, pp. 248-251.

[10] Kuimov D., Minkin, M. Searching of the optimum modes for processing of various origin of cellulose fibers in the electromechanical converter with a discrete secondary part, *AIP Conference Proceedings*, 2019.

Parametric study of the effect of hydrodynamic cavitation on cellulose in an electromechanical converter with a discrete secondary part

D N Kuimov [1,*], **M S Minkin** [2]

[1]Don State Technical University, Department of scientific research, 344000 Rostov-on-Don, Russia
[2]Don State Technical University, Department of automation of production processes, 344000 Rostov-on-Don, Russia

*e-mail: kuimov_d@list.ru

Abstract. Hydrodynamic cavitation is a very effective destructive effect on cellulose fiber. This paper presents a study of the effective parameters of the hydrodynamic cavitation process implemented by an Electromechanical Converter with a discrete secondary part for effective pretreatment of cellulose fiber.

1. Introduction

The availability of an unlimited and renewable natural resource makes pulp and its derivatives one of the most important products for use in various fields and fields of industry. The high importance of the product determines the high relevance of research projects aimed at improving the properties and quality of cellulosic materials, as well as reducing production costs.

To date, the scientific literature has presented many studies on the effects of ultrasonic cavitation on cellulose fiber [1, 2]. However, if hydrodynamic cavitation is used in the process of breaking the cellulose fiber, there is only fragmentary information about the effectiveness of this energy effect. At the same time, in recent years, a direction has been actively developing to study the capabilities of nano-sized cellulose particles, the use of which will allow creating new materials with unique properties. In the development of new processes for the production of nano-sized cellulose particles, technologies are of particular value that allow controlling the processing process with the possibility of changing the properties and parameters obtained by part of the pulp.

The existing processes of grinding cellulosic fibers to nanosized particles are in most cases associated with significant energy costs. Among the methods already developed, cellulosic fibers are ground by mechanical influences such as crushing, cutting, impact, etc.

To carry out grinding processes, a large number of structural versions of machines already exist. At the same time, the search for new unique and innovative technologies for grinding cellulose fiber does not stop. One of these promising technological directions is the effect of hydrodynamic cavitation on cellulose fiber. The destruction of the material in this case occurs under the influence of shock waves and microjets of liquid. These processes are used in industry, however, these phenomena are not sufficiently studied. Little equipment has been developed for efficient grinding processes.

Currently, studies on the effects of hydrodynamic cavitation on cellulosic fiber are very limited. This research aims to enhance knowledge in this field.

2. Main part

In this study, an electromechanical converter with a discrete secondary part (ECDSP) [3-5] was used to generate hydrodynamic cavitation in the pre-treatment of cellulose fiber. Unlike diaphragms, ECDSP is able to create a cavitation cloud under certain conditions throughout the entire volume of the working chamber, significantly accelerating the processing process, while increasing the energy efficiency of the process.

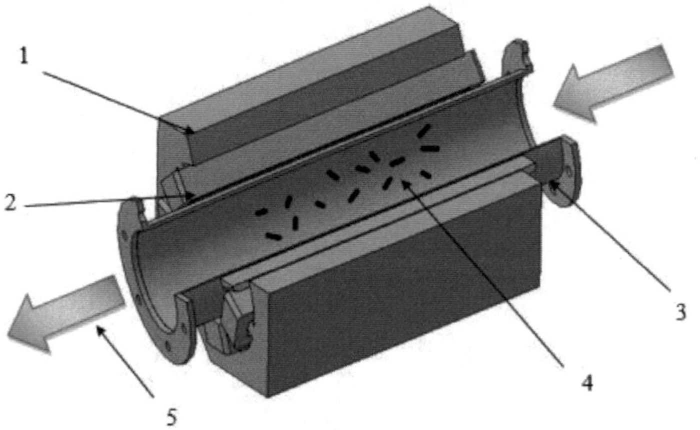

Figure 1. Typical design of electromechanical converter with secondary movable part. 1 - inductor; 2 - 3-phase distributed winding; 3 shows a pipe of non-magnetic material; 4 shows a set of ferromagnetic elements; 5 - treated raw material flow

In addition to hydrodynamic cavitation, the treatment in ECDSP is accompanied by a radome of additional energy effects, including the impact of elements, magnetostriction and friction.

In early experimental studies, the effectiveness of the process of breaking cellulose fibers by hydrodynamic cavitation in ECDSP was confirmed. Examination of the fractional composition of the mass treated in the plant showed that the grinding was carried out by fibrillation and splitting of fibers in length with their slight shortening, which is essential for the quality of the resulting mass. Hydrodynamic cavitation treatment of the mass in ECDSP increases the rate of dewatering and reduces its moisture retention. By using ECDSP, the amount of fibers precipitated several times decreases. In this scientific study, the slope is used to study the most optimal parameters of the process of grinding cellulosic fiber of hydrodynamic cavitation in ECDSP. According to the present invention, the efficiency of the plant depends simultaneously on several factors:

- maximum speed of ferromagnetic element movement;
- temperature of the treated mass;
- residence time of the treated mass.

2.1. *Evaluation of cellulosic fiber treatment duration by hydrodynamic cavitation*

Evaluation of quality of pulp fiber grinding process in ECDSP repeats similar studies performed in [6]. This experimental study is necessary to confirm the previously obtained results.

Results of tests of pulp fiber grinding process in electromechanical transformer with discrete secondary part are given in table 1.

Table 1. Treatment duration dependence on pulp quality

Processing time, min	Average fiber length, mm
10	2,8
20	2,4
30	1,91
40	1,81
60	1,75
80	1,7

2.2. Evaluation of the speed of the ferromagnetic element

Despite the number of research works, there is currently no unified concept of mathematical modeling, there is no way to describe and evaluate the process of movement of the discrete secondary part. The nature of the process of movement of ferromagnetic elements in the device depends on many factors that affect the course of processes and the final result. So, as part of the study, it is necessary to determine at what effective speed the ferromagnetic elements must move in order to ensure the performance of the process.

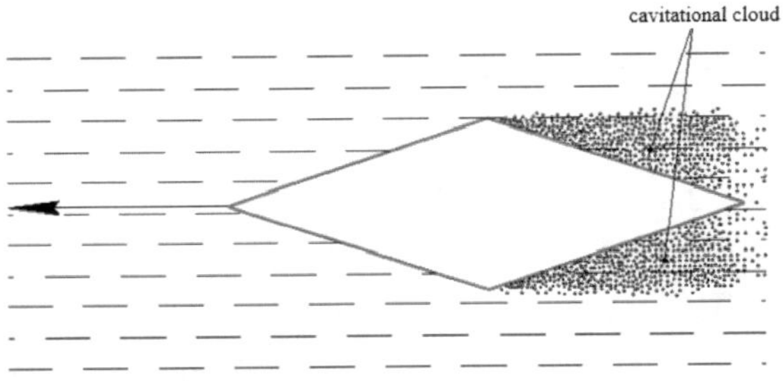

Figure 2. Single ferromagnetic element with supercavitating profile

Cavitation bubbles are formed where the liquid pressure is below a certain critical value (saturated vapor pressure). At the shape of the flowing cavitating element (figure 2), cavitation will occur at a certain value of the dimensionless parameter [7-9]:

$$\chi = 2\frac{p - p_{н.п.}}{\rho \upsilon^2}, \qquad (1)$$

where p – is the hydrostatic pressure of the incoming flow, p_{ss} – the pressure of saturated steam, - the density of the treated liquid medium, ρ – the velocity of the element relative to the treated liquid medium.

Depending on the size of the dimensionless cavitation criterion, 4 types of flow can be distinguished:

$\chi>1$- lack of a cavitation;

$\chi\approx1$- cavitational;

$\chi<1$- developed cavitation;

$\chi<<1$- supercavitational.

The presented four modes are characterized by different cavitation intensity: during the pre-cavitation mode, the continuity of the treated liquid medium is not disturbed, the k coefficient is greater than 1; cavitation mode shows cavitation origin, which is characterized by cavitation vapor-gas bubbles in liquid medium, which saturate liquid medium, coefficient χ is taken equal to 1; operation in developed cavitation mode is connected with approach of cavitation (water-gas) medium resistance to air environment impedance, coefficient χ is less than 1; in the supercavitation mode, cavitation bubbles reach critical dimensions, resulting in subsequent collapse, the coefficient χ is much less than 1.

During operation, a large number of ferromagnetic elements are involved in the working chamber. With small pressure drops, a cavitation cloud will occur only after the first row of bodies. In this case, according to experimental data, the beginning of the formation of a cavitation cloud behind the first row of moving ferromagnetic elements occurs with a dimensionless coefficient of 1, while such a value is not enough to form a cavitation cloud behind the second and subsequent row of moving ferromagnetic elements. If the dimensionless coefficient is less than 1 in the processed raw material, a continuous cavitation cloud occurs throughout the entire volume of the working chamber.

The quality of the treated material was determined by sampling as part of the experimental studies carried out at ECDSP. Treatment was performed for each test sample for 30 minutes. The speed of the ferromagnetic elements was provided by the required current density in the inductor coils and kept constant throughout the experiment [10]. After the corresponding time, the treated material was replaced with a new one and the experiment was carried out again with a new speed of movement of the secondary discrete part. As a result, 5 velocities from 1 to 5 m/s were investigated.

Table 2. Relationship of ferromagnetic element speed to pulp grinding efficiency

Speed of ferromagnetic element movement, m/s	Average fiber length, mm
1	5
2	4,7
3	3,5
4	1,6
5	1,4

Analysis of the obtained experimental data allows to conclude that with the speed of the ferromagnetic element less than 4 m/s, hydrodynamic cavitation does not occur in the treated medium. In this case, the breakdown of the cellulosic fiber occurs by mechanical action, including collisions of a large set of ferromagnetic elements and resulting friction.

2.3. *Temperature estimation of treated raw materials*

According to the present invention, the occurrence of cavitation depends on the speed of movement of the elements under the influence of an external electromagnetic field. However, in terms of the effect of dissolved gas and vapor pressure, the treatment temperature was examined for 30 °C, 50 °C and 70 °C. The duration of treatment for all temperatures was 30 minutes.

Treatment in each case was carried out for 30 minutes. After the corresponding time, the treated material was replaced with a new one and the experiment was carried out again with a new speed of movement of the secondary discrete part. As a result, a number of experimental studies were carried out. Averaged data was included in the final table.

Table 3. Temperature dependence of treated raw materials on pulp grinding efficiency

Temperature of the processed raw materials, °C	Average fiber length, mm
30	2,46
50	1,52
70	1,61

Analysis of the obtained results made it possible to determine that the intensity of cavitation is affected by the amount of gas in the liquid. If the gas in the treated material is in excess, the individual cavitation cavities begin to collide and merge. A cushioning effect is created such that the impact when slammed is absorbed by other bubbles, and the impact is reduced. Thus, the optimum temperature for the pulp grinding process was about 50 °C.

3. Conclusion

A set of experimental studies confirmed the hypothesis about the effect of the speed of movement of the ferromagnetic element and the temperature of the processed raw materials on the efficiency of processing cellulose suspension with an intensive effect of hydrodynamic cavitation. According to the results presented in the study, it can be concluded that the use of optimal parameters of the temperature of the processed raw materials and the necessary speed of movement of ferromagnetic elements provides a better grinding of cellulose with a shorter treatment time. Optimum parameters of treatment are speed of movement of ferromagnetic element not less than 4 m/s and temperature of processed raw material 50°C.

The reported study was funded by RFBR according to the research project № 18-29-18064

References

[1] Gavrilova A.S., Manayenkov O.V., Filatova A. E., Investigation of the effect of ultrasound on microcrystalline cellulose, Bulletin of Tver state technical university, 2015, pp. 60-66.

[2] Sarymsakov A. A., Baltayeva M. M., Nabiev D. S., Rashidova S. sh., Yugay S. M. Dispersed microcrystalline cellulose and hydrogels based on it *Chemistry of plant raw materials*. 2004, №2, pp. 11-16.

[3] Kuimov, D.N., Pavlenko, A.V., Design of the inductor as knot of the electromechanical converter with a secondary discrete part, Lecture Notes in Mechanical Engineering, 2019, pp. 2393-2399.

[4] Kuimov D., Pavlenko A., Minkin M., Optimization of parameters of the electromechanical converter with a discrete secondary part, *MATEC Web of Conferences*, 2017.

[5] Kuimov. D, Minkin M. The electromechanical converter in the systems of desulfurisation of crude oil, *MATEC Web of Conferences*, 2017.

[6] Lukyanov A., Onoiko T., Cellulose crushing: The experiment in the electromechanical transformer with a discrete secondary part, E3S Web of Conferences, 2019.

[7] Yslamidinov A.V. Formation of cavitation bubbles when a water jet passes through the Laval nozzle, International journal of applied and fundamental research, 2016, №. 7, (part 5), pp. 776-778.

[8] Yudaev V. F. hydro-Mechanical processes in rotary apparatuses with modulation of the flow through section of the processed medium, Theor.fundamentals of chemical technology, 1994, Vol. 28, No. 6, pp. 581-590.

[9] Promtov, M. A. Pulsating devices of rotary type: theory and practice. - Moscow: Mashinostroenie-1. - 2001. - 260 p.

[10] Kuimov D., Minkin, M., Searching of the optimum modes for processing of various origin of cellulose fibers in the electromechanical converter with a discrete secondary part, *AIP Conference Proceedings*, 2019.

Modeling and analysis of energy efficiency of methods for maintaining temperature conditions in microbioreactors

V A Chistyakov, A D Lukyanov*, S V Chapek, D Yu Donskoy and O I Katin

Don State Technical University, Rostov-on-Don, 344000, Russia

*e-mail: alexlukjanov1998@gmail.com

Abstract. A mathematical model of maintaining the thermal conditions of microreactors used in biological, medical and other experiments that require stabilization or temperature control software is considered. Several heating modes are considered: using a liquid heating jacket, heating from below through the bottom, heating from the side using a flexible silicone heater. Simulated heat loss and dynamic modes of temperature control.

Introduction

Providing the required temperature conditions during the operation of reactor vessels during biological, chemical or medical experiments is a guarantee of valid and repeatable results. As a rule, various types of heaters are used to ensure the set temperatures: heating tiles, shirts, or flexible heaters fixed to the wall of the reactor.

The aim of the work is to select the optimal quality control structure for energy and energy efficiency to maintain the given temperature of the microreactor, which is planned to be used when conducting experiments in the framework of SHIME type models [1].

1. Methods and models

For research, we will use methods of mathematical modeling of the object using ordinary differential equations (ODE) [2], computer simulation [3] and verification of the results obtained on the experimental layout of the microreactor temperature stabilization system.

1.1 *Mathematical model of a microreactor based on a glass container, basic version*

Consider a cylindrical tank with known geometric and physical parameters, filled with liquid to a level h (figure 1).

The mathematical model of the thermal regime of the microreactor is created using the method of storage and flows (for example, [4]) and in the basic version (only for the microreactor itself) has the form:

$$\frac{d\vartheta_h}{dt} = \frac{1}{\rho_h \cdot V_h \cdot c_h}\left(G - S\left(\alpha_0 + \alpha_1 v\right)\left(\vartheta_h - \vartheta_g\right)\right)$$

$$\frac{d\vartheta_g}{dt} = \frac{1}{\rho_g \cdot V_g \cdot c_g}\left(S\left(\alpha_0 + \alpha_1 v_h\right)\left(\vartheta_h - \vartheta_g\right) - S\left(\alpha_2 + \alpha_3 v_{ext}\right)\left(\vartheta_g - \vartheta_{ext}\right)\right) \quad (1)$$

The model parameters are shown in Table 1. The values of the physical constants are taken from [5].

Table 1. The model (1) parameters

Designation	Dimension	Value	Note
b	m	0.06	Base diameter
h	m	0.08	Chyme level height
d	m	0.001	Wall thickness
ρ_h	kg/m^3	1000	Chyme density
c_h	J/(kg*K)	4200	Heat capacity of chyme
α_0		450	stat. part of water-to-surface heat transfer coefficient
α_1		$2100\sqrt{v}$	dyn. part of water-to-surface heat transfer coefficient
α_2		5,6	stat. part of air-to-surface heat transfer coefficient
α_3		4*v	dyn. part of coefficient water-to-surface heat transfer
λ	W/(m*K)	1.2	Thermal conductivity of Pyrex glass
ρ_g	kg/m^3	2230	Pyrex glass density
c_g	J/(kg*K)	830	Heat capacity of Pyrex glass
V_g	m^3	$\left(\pi \cdot b^2 / 4 + 2 \cdot \pi \cdot b \cdot h\right) \cdot d$	Pyrex glass volume
V_h	m^3	$\pi \cdot h \cdot b^2 / 4$	Chyme volume
S_1	m^2	$2\pi \cdot b \cdot h$	The contact area of the chyme with the walls of the glass
S_2	m^2	$\pi \cdot b^2 / 4$	The contact area of the chyme with the bottom of the glass
v_h	m/c	0.01	The speed of the chyme along the wall
v_g	m/c	0.01	Air velocity along the wall
G	W	5	Heat flow from heater, max.
ϑ_{ext}	°C		Ambient temperature
ϑ_h	°C		Chyme temperature

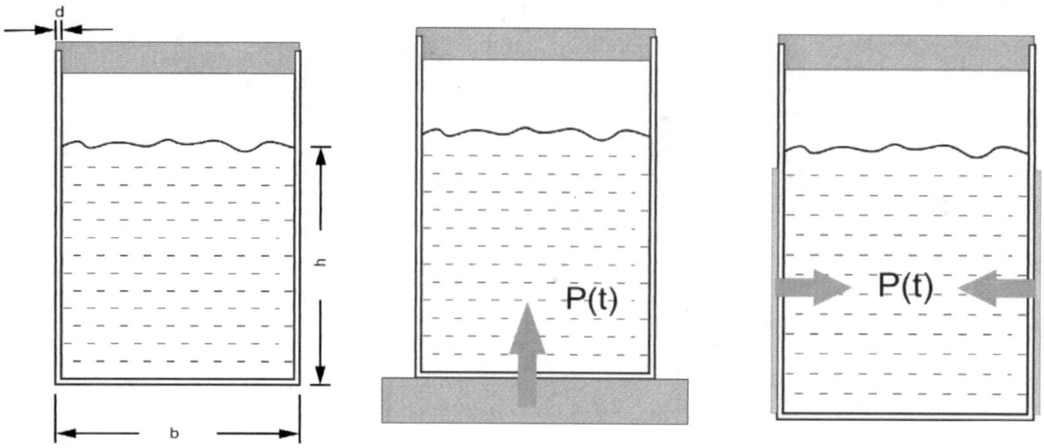

Figure 1. Objects of modeling. “a” – model with inherent heating (1), “b” – model with bottom heating (2), “c” – model with side heating (3)

The model of the dynamic system (1) has the following structure (figure 2)

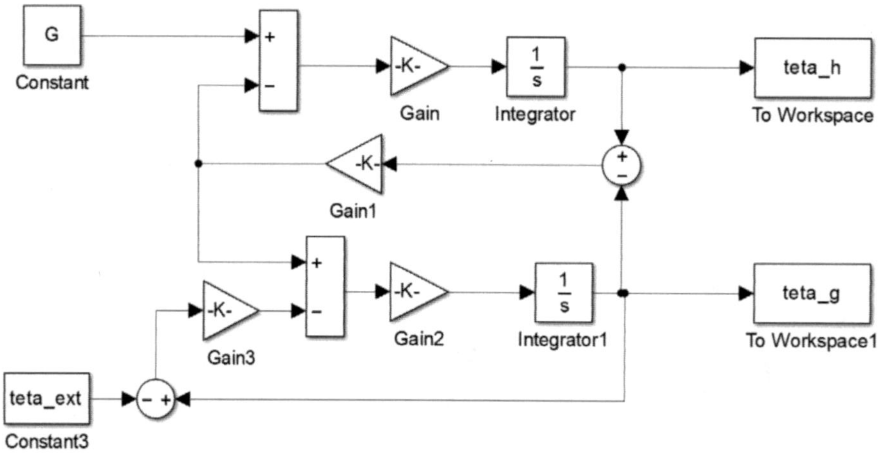

Figure 2. The structure of the model of microreactor with internal heating.

1.2. *Mathematical model of a microreactor based on a glass tank, heated from bottom*

Consider the model of a microreactor heated from below using a classic heater with a magnetic stirrer. In this case, the dynamics of heating will change due to the appearance of another inertial system.

$$\frac{d\vartheta_h}{dt}=\frac{1}{\rho_h\cdot V_h\cdot c_h}\left(S_p\lambda\left(\vartheta_p-\vartheta_h\right)-S\left(\alpha_0+\alpha_1 v\right)\left(\vartheta_h-\vartheta_g\right)\right)$$

$$\frac{d\vartheta_g}{dt}=\frac{1}{\rho_g\cdot V_g\cdot c_g}\left(S\left(\alpha_0+\alpha_1 v_h\right)\left(\vartheta_h-\vartheta_g\right)-S\left(\alpha_2+\alpha_3 v_{ext}\right)\left(\vartheta_g-\vartheta_{ext}\right)\right) \quad (2)$$

$$\frac{d\vartheta_p}{dt}=\frac{1}{m_p\cdot c_p}\left(G-S_2\lambda_p\left(\vartheta_p-\vartheta_h\right)\right)$$

Additional parameters corresponding to the characteristics of the heater are given in Table. 2

Table 2. The model (2) additional parameters

Designation	Dimension	Value	Note
m_p	kg	0.3	Heater mass
c_p	J/(kg*K)	900	Heat capacity of heater (Al)
λ_p	W/(m*K)	500	Coeff. heat transfer from the heater to glass

The model of the dynamic system (2) has the following structure (figure 3)

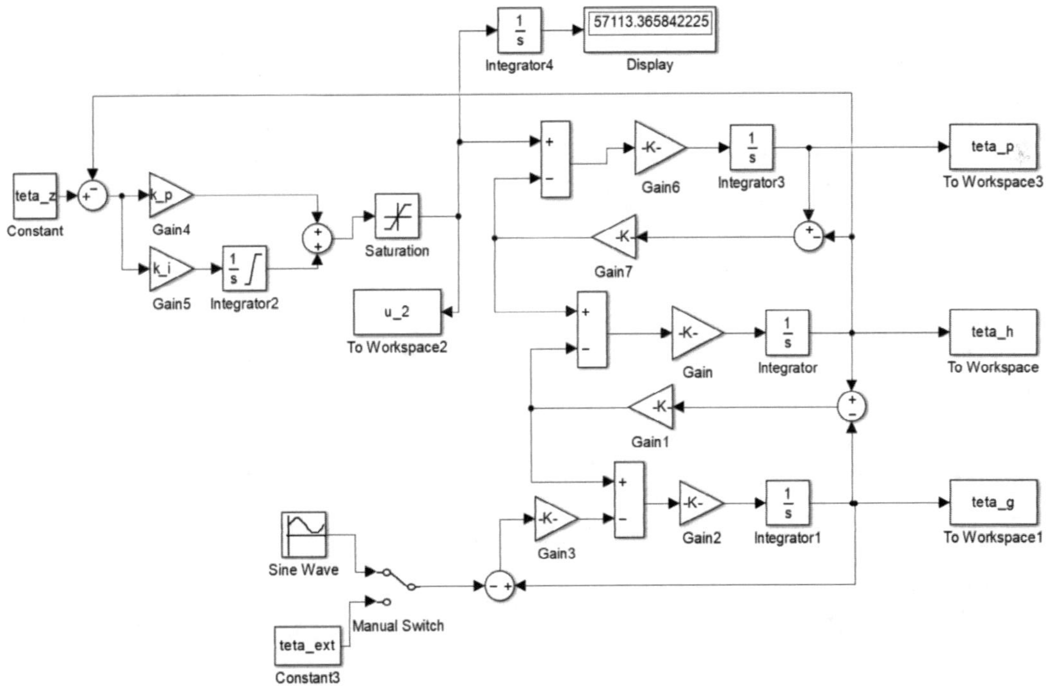

Figure 3. The structure of the model (2) of microreactor with external (bottom) heating

1.3. *Mathematical model of a microreactor based on a glass tank, side heating*

For a silicone film heater located on the side wall of a microrector, the model takes the following form:

$$\frac{d\vartheta_h}{dt} = \frac{1}{\rho_h \cdot V_h \cdot c_h} S_1\alpha_0\left(\vartheta_h - \vartheta_g\right)$$

$$\frac{d\vartheta_g}{dt} = \frac{1}{\rho_g \cdot V_g \cdot c_g}\left(S_1\lambda_p\left(\vartheta_p - \vartheta_g\right) - S_2\lambda_p\left(\vartheta_h - \vartheta_g\right)\right) \tag{3}$$

$$\frac{d\vartheta_{p1}}{dt} = \frac{1}{m_{p1} \cdot c_{p1}}\left(G - S_1\lambda_p\left(\vartheta_p - \vartheta_g\right) - S_1\alpha_2\left(\vartheta_g - \vartheta_{ext}\right)\right)$$

Additional parameters corresponding to the characteristics of the heater are given in Table. 3

Table 3. The model (3) additional parameters

designation	dimension	value	note
m_p	kg	0.03	Heater mass
c_p	J/(kg*K)	1,420	Heat capacity of heater (Al)
λ_p	W/(m*K)	500	Coeff. heat transfer from the heater to glass

The model of the dynamic system (3) has the following structure (figure 4)

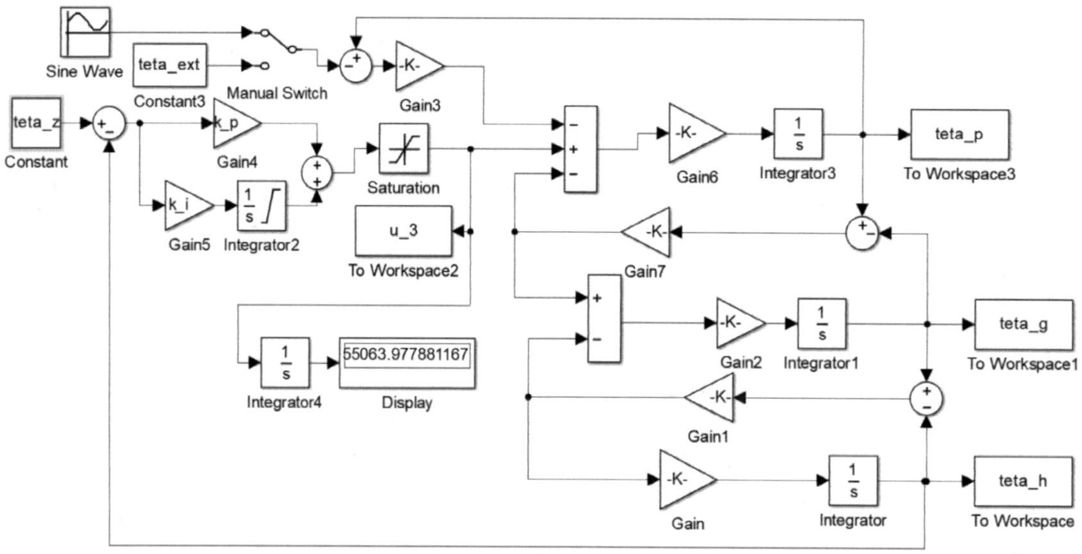

Figure 4. The structure of the model (3) of microreactor with external (side) heating

In this case, the mass of the heater is significantly reduced, reducing its inertia, and the contact area of the heater with the capacity (and chyme) increases.

2. Results of simulations

A computer study of the dynamics of a microreactor based on the above mathematical models was carried out in the mathematical package Matlab.

The simulation result of heating the microreactor with an abstract heater emitting 5 Watts of power inside the chyme is shown in figure 5

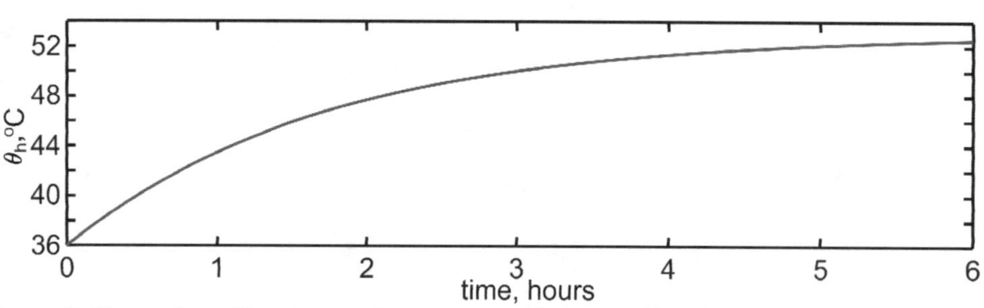

Figure 5. Dynamics of heating a microreactor with internal heating (model (1)) at P = 5 W

Adding a PI-controller to the model allows you to provide a given temperature when the temperature of the environment with an amplitude of 2 °C. Obtained results shown in fig. 6

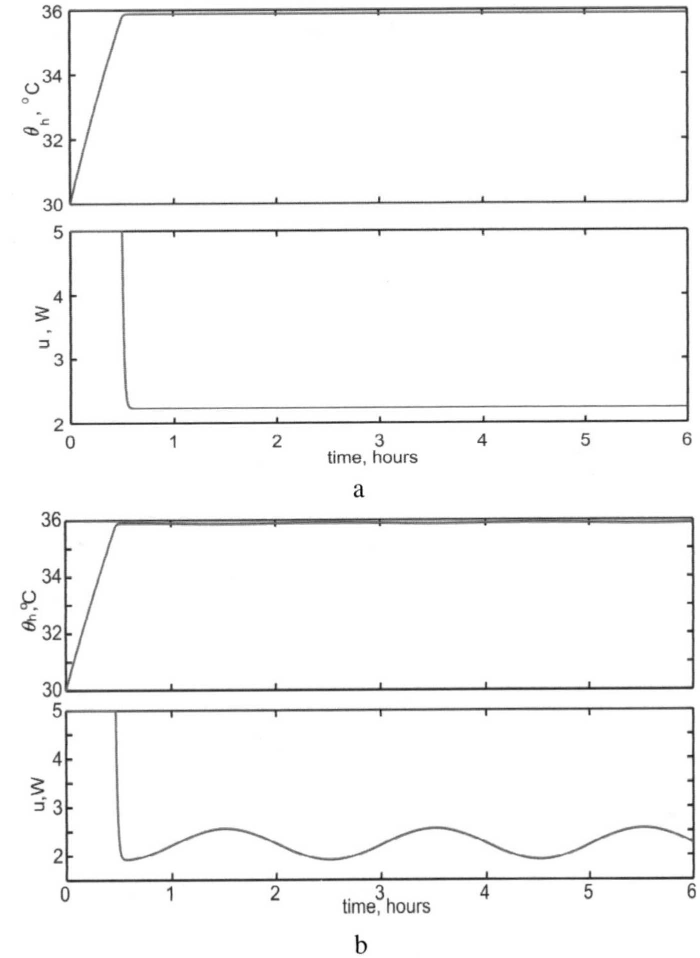

Figure 6. Dynamics of the temperature of a microreactor (model (1)) with a temperature of 36 °C: "a" for a constantly external temperature, "b" for a sinusoidally changing temperature

For PI-controller parameters $k_p = 20$, $k_i = 0.1$, the following graphs were obtained for model (2) (figure 7).

The simulation results of the dynamics of heating the microreactor on the side (model (3)) with the same parameters of the PI controller are shown in figure 8. By reducing the inertia of the heater, the dynamics of the system changes positively.

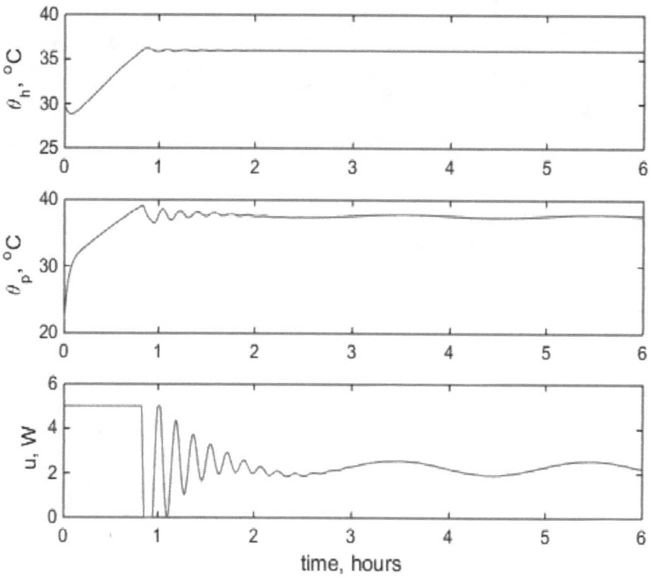

Figure 7. Dynamics of the temperature of a microreactor (model (2)) with a temperature of 36 °C

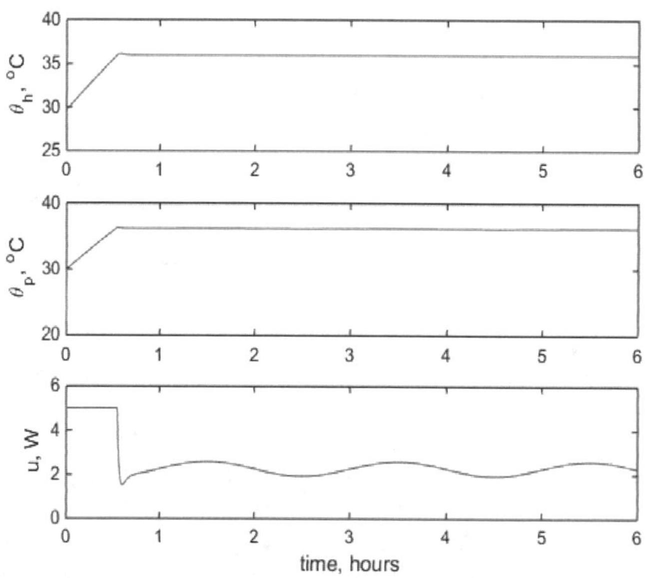

Figure 8. Dynamics of the temperature of a microreactor (model (3)) with a temperature of 36 °C

3. Experimental layout

To compare the results of computer simulation with the heating of a real microreactor, an experimental model was developed that included a laboratory machine as a microreactor, a microprocessor control system, and a laptop computer for recording the results (figure 9). The structure of the microprocessor control system is shown in figure 10. As the control controller, Arduino UNO was used, in the program of which a digital controller and a PWM controller for controlling the heater were implemented. To measure the chyme temperature and external temperature, digital thermometers of the DS18B20 type [6] with a 1-Wire interface were used. Some decisions were borrowed from [7].

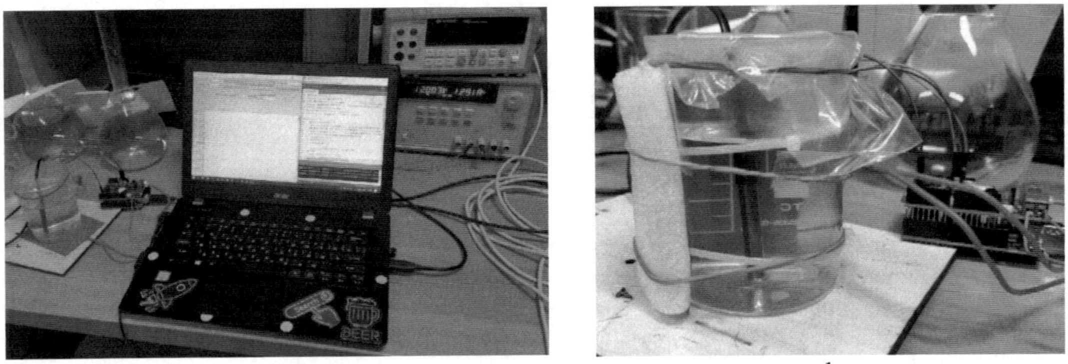

a b

Figure 9. Appearance of the laboratory layout. "a" is the system as a whole, "b" is a microreactor simulator

To ensure heating, a flexible heater in a silicone case was used, covering approximately 50% of the microreactor wall. Heater power is 5 watts.

During the experiment, the measurement results (chyme temperature, ambient temperature, current consumed by the heater) were transferred via a serial interface to a portable computer, and saved to a file. The measurement results are shown in figure 11 against the background of the results of simulation modeling of model (3) under the corresponding initial and external conditions.

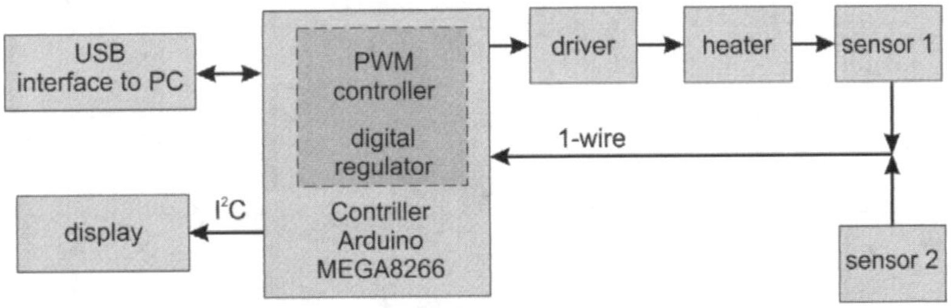

Figure 10. The structure of the microprocessor control system

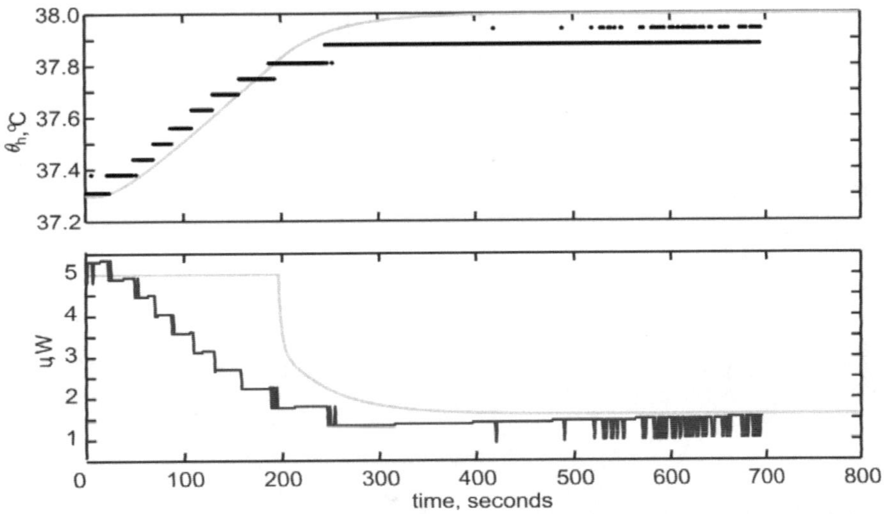

Figure 11. The measurement results (points & blue line) against the background of the results of simulation modeling of model (3) (green line)

IOP Conf. Series: Materials Science and Engineering **900** (2020) 012015 doi:10.1088/1757-899X/900/1/012015

4. The discussion of the results

Three mathematical models of heating a microreactor were developed and investigated by computer simulation methods: with ideal heating in the volume of the chyme, with heating from the bottom and heating from the side of the side wall. A digital controller was included in the model, which was supposed to provide stabilization of the chyme temperature at a given level. The simulation results presented in figure 11 showed that the system with side heating is closer in characteristics to the ideal model, and on this basis, it can be recommended for implementation.

During the study of the models, the total energy spent on maintaining the set temperature in each of the cases was estimated, it is given in table. 4 and illustrated in figure 12.

Table 4. The total energy spent of different models

Models	Energy, J
Model (1)	53419.66
Model (2)	56623.92
Model (3)	55063.93

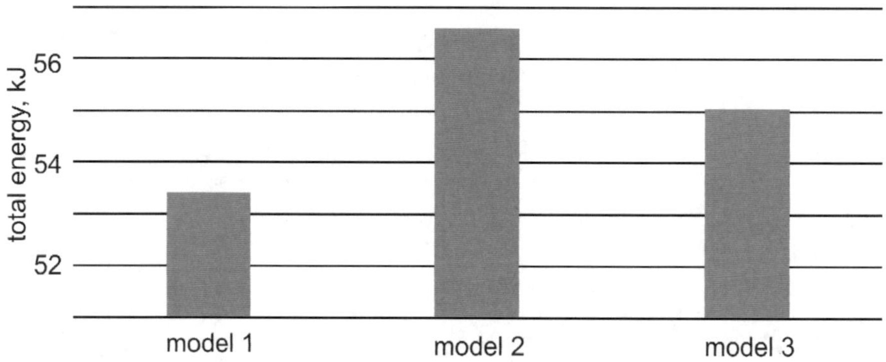

Figure 12. Histogram of total energy spent of different models

Based on the data obtained, it can be argued that according to the energy criterion, a model with side heating (model (3)) is preferable to a model with heating through the bottom (model (2)). Figure 13 shows comparative graphs of the control action on the model, with a sinusoidal change in external temperature. In this case, model (3) (red line) is also closer to the ideal model (1) (green line).

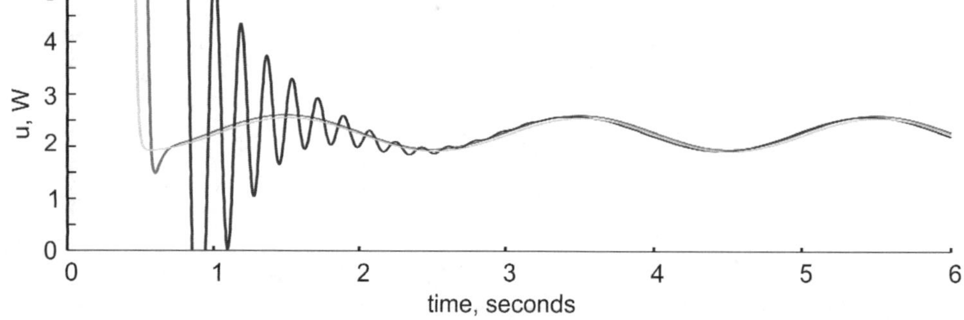

Figure 13. Comparative graphs of the control action on the model, with a sinusoidal change in external temperature

Conclusions

In order to verify the system's operability in practice, an experiment was conducted to heat the chyme simulator (water) using a flexible heater, and the results (figure 11) showed good qualitative and good quantitative agreement with the simulation results. Some discrepancy in the operation of the digital controller is possible due to its simplified implementation in the microcontroller, compared to Matlab. The results obtained made it possible to justify the choice of lateral heating as the most suitable for implementation in the system of maintaining the temperature regime of the microreactor for biological and medical experiments.

Acknowledgments

The research was supported by the Grant Russian Ministry of science & higher education No. 075-15-2019-1880.

References

[1] Van de Wiele T., Van den Abbeele P., Ossieur W., Possemiers S., Marzorati M. (2015) The Simulator of the Human Intestinal Microbial Ecosystem (SHIME®). In: Verhoeckx K. et al. (eds) The Impact of Food Bioactives on Health. Springer, Cham

[2] Karthikeyan N., Jayaraja A. Application of First Order differential Equations to Heat Transfer Analysis in solid // International Journal of Engineering and Innovative Technology (IJEIT), Volume 5, Issue 8, P. 5-8.

[3] Modeling Thermal Liquid Systems // Matlab documentation. The MathWorks, Inc. URL: https://www.mathworks.com/help/physmod/simscape/ug/thermal-liquid-modeling-workflow.html

[4] Susana Pelayo Díaz. Modelling and Simulation of an Industrial Steam Boiler with Ecosimpro // 1st Meeting of EcosimPro Users, UNED, Madrid, 3-4 May 2001, P. 1-10

[5] Horst Kuchling. Taschenbuch der Physik // 21 aktualisierte Auflage. 10/2014, 711 Seiten. Zweifarbig

[6] DS18B20 Programmable Resolution 1-Wire Digital Thermometer // Maxim Integrated Products, Inc. URL: https://datasheets.maximintegrated.com/en/ds/DS18B20.pdf

[7] Konev, S.A., Vernezi, M.A., Donskoy, D.Y., Lukyanov, A.D. The development of a mobile monitoring system for agricultural object. Volume 132 (2017) MATEC Web Conf., 132 (2017) 04026

IOP Conf. Series: Materials Science and Engineering **900** (2020) 012016 doi:10.1088/1757-899X/900/1/012016

5G tool for evaluation and comparision of energy efficiency of mobile radio channel using second-order statistics

N Stefanović[1], A Kar[2] and V Mladenović[1,*]

[1]Faculty of Technical Sciences in Čačak, University of Kragujevac, Svetog Save 65, Čačak, Serbia
[2]Indian Institute of Information Technology, Design and Manufacturing, Kancheepuram, Chennai, India.

*e-mail: vladimir.mladenovic@ftn.kg.ac.rs

Abstract. In order to increase energy efficiency over transmissions channels, common approach for optimization tasks is by means of channel's second-order statistics. Actual channel modeling tools for 5G networks end with channel's first-order statistics, although these metrics are not sufficient when channel conditions are rapidly changing, either in time, frequency or space. In this paper, we establish a tool for evaluation and comparision of energy efficiency of mobile radio channel using its second-order statistics, especially level crossing rate (LCR) and average fade durations (AFD), as they can implicitly pinpoint to transmission configurations that are energy efficient or, as oposit, become a waste of energy. Using both deterministic and stochastic channel modeling, we present results after simulations of Rayleigh channel for narrowband case and further extend it to passband cases, suitable for 5G scenario. We conclude about the energy efficiency of different transmission schemes used by the 5G physical layer observing LCR and AFD values.

1. Introduction

Next generation wireless systems, aiming to support communication in reliable and robust – anytime, anywhere manner, meets challanges to provide best possible performance levels [1] and concurently satisfy defined quality of service (QoS) constraints (eg. packet error rate, packet delay budget) [2], as 5G is claimed to be first QoS driven network [3]. Beside randomly varying channel conditions, due to multiplicity of technologies applied, emerging number of device users and divergence of use cases, one of key challanges is making efficient use of energy resources.Upon this circumstance, question arise how to optimize different connection links in order to save energy as much as possible. Researches have shown that, for some wireless communication link, characterized with one of its functions - channel matrix, it is required to acquire channel's second-order statisticsto provide optimal precoding matrix and maximize energy efficiency of transmission [4]. Second, in a channel affected by varying multipath fading process, first-order statistics, such as probability density function (PDF) and cumulative distributive function (CDF) are not sufficient as they describe only its statical behaviour. To describe process dynamics, one needs higher-order statistics, such as level-crossing rate and average fade duration [5].

Previous observation leads to considering appropriate basis for development, optimization and test of transmission system through channel modelling. Latest requirements in channel modelling are to

IOP Conf. Series: Materials Science and Engineering **900** (2020) 012016 doi:10.1088/1757-899X/900/1/012016

accurately capture space-time-frequency characterstics of the mobile radio channel in a variety of scenarios and specificaly first and second-order statistics, which includes the LCR and AFD [6]. This modeling is achieved by deterministic, stochastic or hybrid approach. Models can be derived mathematicaly or from measurements in a physical, real-world channel, as illustrated on figure 1.

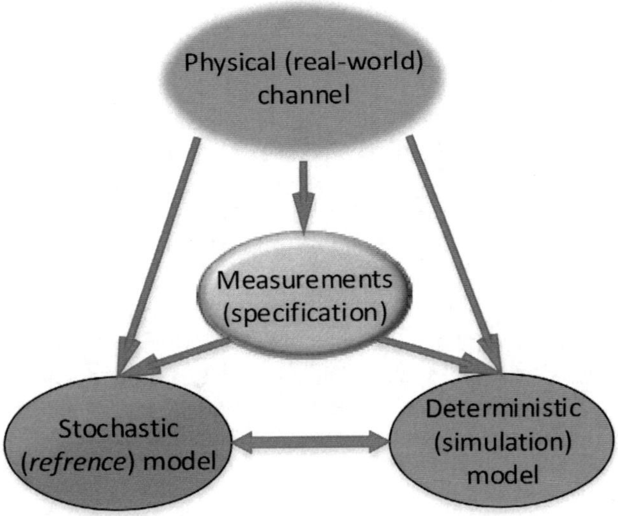

Figure 1. Relations between different channel modeling aspects. Measurements of physical, real-world channel support validity of other models

While deterministic models usualy employ ray tracing technique – computationaly intensive, inherently less accurate, but reliable, stochastic or statistical models are mathematicaly derived or estimated upon measurements of a physical, real-world channel and – designed for fast, system level simulations.

Comprehensive (but not final) list of actual tools for 5G channel modeling is stated in [7]. All tools, as the author's best knowledge, end with analysis of first-order statistics for 5G use cases, which is not enough knowledge to predict dynamic behaviour of channel. In our paper, we establish a tool for evaluation of mobile radio channel using its second-order statistics and make adequate conclusions of energy efficiency for each use case analyzed. Our motivation goal was the fact that optimized transmission can drasticaly save energy consumption. For presentation sake, urban environment was supposed where no line-of-sight between transmitter and receiver exists and multipath propagation is immanent. Hence, receiver is moving terminal, involving effects like Doppler shift and signal phase variations resulting in signal's complex envelope (CE) fluctuating and fading below some receiver's sensitivity level. Apart from other, with our tool we were able to examine important metrics, as LCR and AFD and their behaviour for different channel parameters. In addition, we extended our observation to wideband model and put it in a 5G transmission schemes framework.

The rest of the paper is organized as follows. In the section 2, we give short review of other state-of-the-art tools for 5G channel modeling with discussion on second-order statistics usage. In section 3, starting from well known Clarke's deterministic model in Rayleigh channel, we put short description of a tool with observation considering a case with a moving receiver on small travell route section and capturing CE to evaluate its statistics. Hereafter, we put our simulation setup in the context of 5G transmission schemes, known as numerologies and finaly, in section 4, present results concluding on benefits and further work possibilities of presented tool.

2. On channel model tools with second-order statistics for 5G use cases

2.1. Channel models for 5G use cases

Sublimated list of provided in 3GPP's study gives an comprehensive overview of actual channel model tools for next generation network evaluation and design. First comes 3GPP's NR (New Radio) channel model in the form of TDL and CDL channels and implemented in Matlab's 5G Toolbox, bringing all relevant setup for testing, signal generation and analysis, as map-based propagation simulating scenarios [8]. In QuaDRiGa (QUAsi Deterministic RadIo channel GenerAtor) [9], geometry-based stochastic channel model is used of multipl-input-multiple-output (MIMO) radio links providing all relevant parameters as absolute path delays, Doppler spread, channel impulse response (CIR), etc. Based on this model, Quamcom proposed simulator [10] for further research and development of 5G NR physical layer [11]. Statistical, cluster based, modeling tool as COST 2100 [12], was developed and used for analysis some time before of ITU-R's vision of 2020 and beyond, focusing on multi-user, distributed MIMO and moving Tx-Rx scenarios in the 4G communication system. Program METIS [13], belongs to deterministic types of channel modeling tools, it was intended to complement time-consuming and expensive measurement campaigns for overall 5G scenarios, offering wide range of parameters and scenario's setup, but as previous finalizing with PDF and CDF statistics. At the end to mention, but not as final on the list, is NYUSIM [14] model, built up from extensive propagation measurements carried out during time period of four years and proposed as statistical spatial channel model using time cluster-spatial lobe approach. Designed primarly for mmWave directional channels, as presented, does not cover second-order statistics analysis, although fading behaviour in this range is literally highly dynamic [15].

2.2. Second-order statistics analysis

As for the LCR and AFD statistical metrics from 5G view, not much results are exploited in available papers. One of rare works is done by [16], where different carrier mmWave statistics was presented but no passband behaviour examined, nor energy efficiency of such transmission. Another performance analysis of 5G transmission over fading channels was conducted in [17], [18] again without passband or efficiency performance. In tool proposed through this paper, we extend current analysis with passband cases, suitable for appliance to flexible 5G transmission schemes, called numerologies [19] and related to different subcarrier spacings (SCS) in chosen orthogonal-frequency-division-multiplex (OFDM) waveform.

Before going onward, it is desirable to illustrate these metrics, figure 2, as they show how often signal crosses defined treshold and how long on average the signal remains below that level.

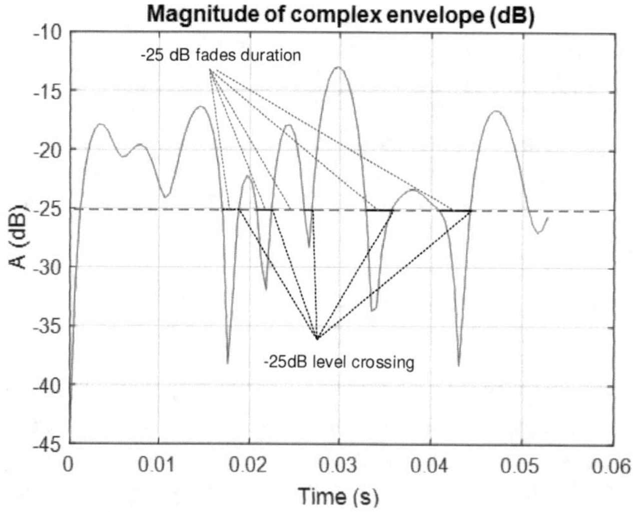

Figure 2. Level-crossing rate and average fade duration definition

Analytical expression for rate that complex envelope r is crossing level L, is derived from its definition:

$$N_L = \int_0^{\infty} \dot{r} p_{r\dot{r}}(L, \dot{r})\, d\dot{r}, \qquad r \geq 0 \tag{1}$$

where $p_{r\dot{r}}(L, \dot{r})$ denotes joint probability density function of stochastic variable $r(t)$ particularized at level L and its derivative $r^{\cdot}(t) = {dr}/{dt}$ at the same time instant. By definition, AFD is the expected value of the length of the time intervals in which stochastic process is below a given signal level L and is calculated by

$$T_{afd} = \frac{P_{CDF}(r < L)}{N_R} \tag{2}$$

and $P_{CDF}(r < L)$ is cumulative distributive function – probability that stochastic process (complex envelope r) is less or equal to L. These metrics, although often hard to express analyticaly, numericaly are not intensive to calculate and measure. Additionaly, we refer to the definition of energy efficiency as maximum number of bits that can be delivered per joule of consumed energy in the network,

$$C_J = \frac{R}{P_\Sigma} \tag{3}$$

where R presents data rate (b/s) and P_Σ total consumed energy. For symplicity, here we overide presenting expressions for single each use case, as it can occupy too much space. Other definitions are rate per energy, or rate per unit cost.

3. Tool description and simulation setup

3.1. Tool description

Real-world multipath/multiscatter propagation scenario is often depicted as on figure 3. Apart from usual presentations, spatial separation is highligted, meaning presence of multiple beams from base stations (denoted as gNB) in the same environment. Mobile user (well known as User equipment - UE) is moving with some constant velocity V, causing an existance of Doppler frequency at each beam i that arrives at the receiver with the angle α_i. Buildings, roofs, towers, high trees, etc, make the environment multiscattering and consequently multipath propagated. Starting from well known Clarke's deterministic model, at this instance we focus on fast variations of signal complex envelope r at receiver point (fast fading) when no direct line-of-sight is available. It is main source of signal drops and degradation in this case (urban environment). For analysis of fast fading separate from slow variations (slow fading due to distance change), one must define *small local* area, where observing route section Δl takes tipicaly 10 – 40 wavelengths (λ) of transmitted signal. Samples of r are taken at some fractions of wavelength F, at each $\Delta x = \lambda/F$ point of UE movement, equivalently every λ/FV second. Regarding to multipath of rays, two effects are considered at the receiver: Doppler and phase variations. For i-th ray, contibution to receiving signal with Doppler shift depends on the angle of arrival $f_{Di} = f_{max} \cos \alpha_i$, where $f_{max}=V/\lambda$ presents maximum Doppler shift. Phase increment of signal in sample point $x[n]$ equals

$$\Delta\phi_n = \frac{2\pi}{\lambda}\, d[n] \tag{4}$$

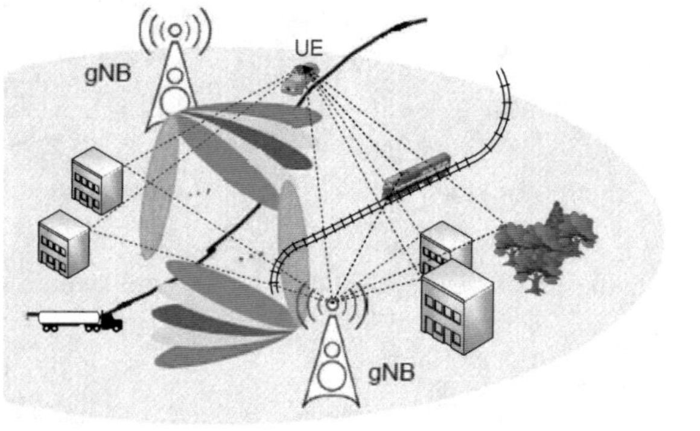

Figure 3. Scenario of 5G multipath propagation environment.

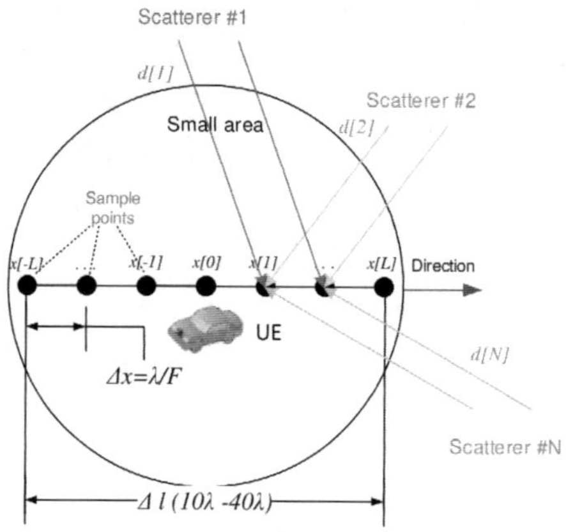

Figure 4. Small local area.

Term $d[n]$ represents path distance that ray travels from base station to receive point. That way, narrowband channel model was built up, leading to deterministic expression of receiving complex envelope term

$$r(t) = a_o e^{-j2\pi \frac{V}{\lambda} t} = a_o\, e^{-j\frac{2\pi}{\lambda}d[n]} \tag{5}$$

where a_0 presents its constant magnitude. Considering small area, it is assumed that rays coming from the same scatterer in each sampling point are parallel, figure 4. Urban environment, on the other hand is often statisticaly described with Rice or Rayleigh distribution, where it is shown that LCR

$$N_L = \sqrt{\frac{\beta}{2\pi}\frac{L}{\sigma_0^2}} e^{-\frac{L^2}{2\sigma_0^2}} = \sqrt{\frac{\beta}{2\pi}} p_r(L), \quad L \geq 0, \tag{6}$$

showing that practicaly LCR depends on acf $p_r(L)$, where σ_0^2 denotes mean power and β is short notation of negative curvature for the autocorrelation function (ACF) at the origin τ =0,

$$\beta = -\frac{d^2}{d^2\tau} R(\tau)\bigg|_{\tau=0} \tag{7}$$

and $R(\tau)$ is ACF of CE. At the same time, ACF is found to be

$$R(\Delta x) = J_0\left(\frac{2\pi}{\lambda}\Delta x\right) \tag{8}$$

with J_0 noted well known Bessel function of first kind and zero order. Note that distance and time are connected here through $\Delta x = V\Delta t = V\tau$. Analogously, for AFD in Rayleigh channel, we can express

$$T_{afd} = \sqrt{\frac{2\pi}{\beta}\frac{\sigma_0^2}{L}}\left(e^{-\frac{L^2}{2\sigma_0^2}} - 1\right), \quad L \geq 0, \tag{9}$$

From the front user interface of our tool, set of parameters can be adjusted, according to small area description, figure 4: distance from the base station, UE velocity, number of scatterers, number of samples (sample points) and measurement sample rate (related to fraction *F*), frequency (wavelength), transmission power and simulation length. In the case shown on figure 5, scatterers are distributed randomly around UE, in some radius which can be arbitrary set. Small area is presented with a red dot or dash, proportional to calculated value of route Δl. On the right side of a view there are several different analysis results displayed within separate tabs: amplitude and phase behaviour of complex signal envelope, Doppler spectrum, first-order statistics including cumulative distributive function (CDF) and autocorrelation function (ACF), second-order statistics including LCR and AFD, as well as random FM.

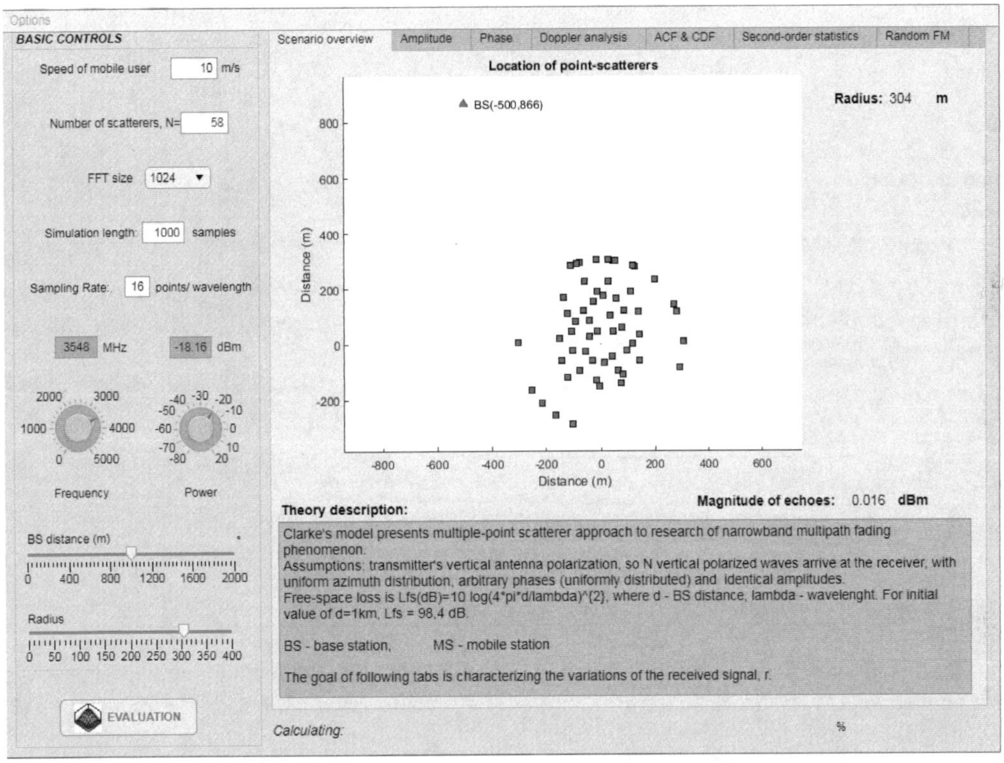

Figure 5. Graphical user interface of 5G tool used for LCR and AFD analysis

3.2. Simulation setup

After description of applied deterministic and stochastic mechanisms, we are ready to set our simulation according to predefined transmission schemes used in 5G communication. These flexible schemes, called numerologies, table 1, are related to slicing available bandwidth to bandwidth parts which can use different predefined SCS combination, according to channel conditions. To make parallel with real state condition, we consider that measurements in 5G are done on base station's specific secondary synchronization signal blocks (SSB) transmitted periodically and separately on single spatial beam. These blocks are 4 OFDM symbol long in time and 240 subcarriers wide in frequency scale.

One more consideration is carrier frequency around which passband signal is existing, called reference point *A*. Two frequency ranges (FR) are adopted in 5G, as shown in table 1 and reference point frequency, say f_A is not fixed and can be different for each block. We started our arbitrary scattering scenario with wideband signal, in FR1, more precisely with a starting frequncy at NR operating in band *n*77, which is 3300-3800 MHz. After receiver detects f_A, orientation is done and receiver knows exact location of a reference point. Therefore, our scanning parameter should be set at the beginning and band will add SCS on it.

IOP Conf. Series: Materials Science and Engineering **900** (2020) 012016 doi:10.1088/1757-899X/900/1/012016

Table 1. Flexible 5G transmission schemes, called numerologies.

Subcarrier spacing – SCS (kHz)	15	30	60	120	240
Frequency range -FR	**1**:410-7125 MHz			**2**:24,25-52,6GHz	
Symbol duration (μs)	66,7	33,3	16,7	8,33	4,17
CP duration (μs)	4,7	2,3	1,2 – normal 4,3 – extended	0,59	0,29
Max. nominal bandwidth (MHz)	50	100	100 – FR **1** 200 – FR **2**	400	400
Maximal FFT size	4096	4096	4096	4096	4096
Symbols/slot	14	14	14 12 – extended	14	14
Symbols/subframe	1	2	4	8	16
Slots/frame	10	20	40	80	160

We choose 3500 MHz, or λ=0,086 m and route we observe is 0,86 – 3,44 m (small area). Depending of numerology applied, bandwidth (240 subcarriers) and signal duration (3 symbols) take values according to table 2.

Table 2. Simulation setup parameters.

SCS (kHz)	15	30	60
Numerology	Case A	Case B	Case C
Symbol duration (μs)	200	100	50
Bandwidth (MHz)	3,6	7,2	14,4

Relating this to energy efficiency, we recall (3) intending to increase throughput, while maintaining constant enaged energy, figure 6. Increasing throughput, means that we are looking for shorter AFD and lower LCR regions.

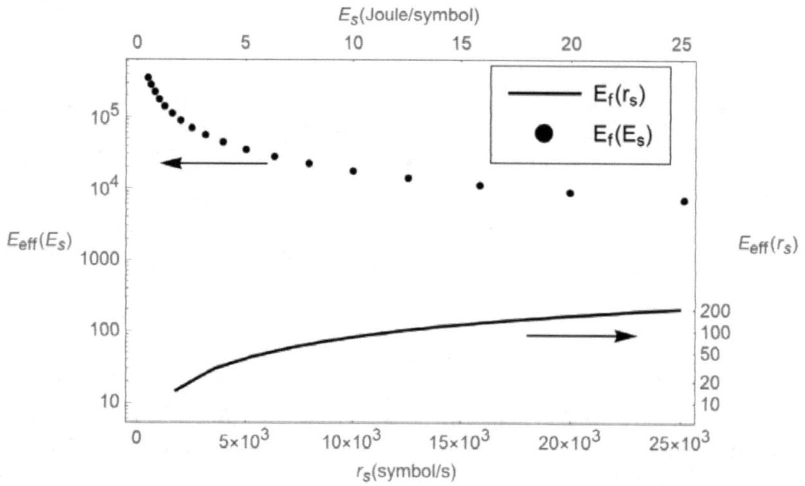

Figure 6. Energy efficiency in OFDM system at constant energy consumed

4. Simulation results

In this session we show and describe Case A. We show LCR and AFD for narrowband signal case (single frequency at 3500 MHz), figure 7. To define small area (40λ) we take 1000 samples with sample rate 25 points per wavelength. Velocity of UE is set to 36 km/h. Most of signal is fluctuating around -3dB with reference to rms value of CE, where it reaches 1 crossing/wavelength (~ 86 mm of route). AFD is normalized to Doppler shift, for reference level set to 0 dB, most of signal fades out, transmission becomes waste of energy.

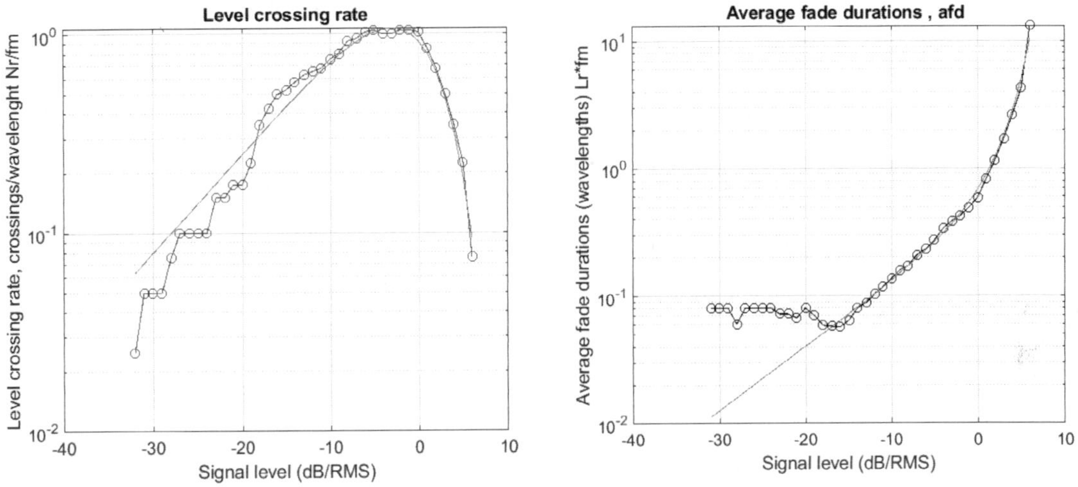

Figure 7. a) Normalized LCR in term of level *L* **b)** Normalized AFD in term of level *L*

Results show good congruence between theoretical (full line) and simulated series (connected dots), validating our channel model for narrowband case.

5. Conclusion

Second-order statistics on channel state variations can help to save energy from trying lossy transmissions. Channel modeling is prerequisite for such optimization and level crossing rate and average fade duration are metrics not intensive to calculate and measure. Present state of tools explicitly, they do not use these metrics for 5G channel modeling. Establishing a tool to evaluate channel's LCR and AFD through signal's passband can improve next generation networks energy efficiency by optimizing flexible transmission schemes used.

Tool presented, yet, leaves wide space for further work and improvements. First, scenario editor can be extended to consider real-map 3D models as in some other tools. Deriving expressions for energy efficiency dependent on LCR/AFD, graphics could be more obvious in terms of optimization. Finaly, mapping QoS requirements directly to physical layer metrics as LCR and AFD, can bring reliable intuition about justification for some network configuration deployment.

Acknowledgments

This paper is supported by the Ministry of Education, Science and Technological Development of the Republic of Serbia, and these results are parts of the Grant No. 451-03-68/2020-14/200132 with University of Kragujevac - Faculty of Technical Sciences Čačak.

IOP Conf. Series: Materials Science and Engineering **900** (2020) 012016 doi:10.1088/1757-899X/900/1/012016

References

[1] ITU-R 2015 Recommendation ITU-R M.2083-0: IMT Vision – Framework and overall objectives of the future development of IMT for 2020 and beyond, (Geneva, International Telecommunication Union)

[2] 3GPP 2020 TS 23.501v16.4.0 : System Architecture for the 5G System (5GS) (www.3gpp.org/ftp/Specs/archive/23_series/23.501/).

[3] Kotkamp M, Pandey A, Raddino D, Roessler A and Stuhfauth R 2019 5G New Radio, Fundamentals, Procedures, Testing Aspects (Munich: Rohde & Schwarz) p 285.

[4] Wu J Rangan S and Zhang H 2013 Green Communications: Theoretical Fundamentals, Algorithms and Applications (Boca Raton: CRC Press, Taylor & Francis Group) pp 520-547

[5] Anjos A Marins T, Souza R and Yacoub M 2018 Higher order statistics for the α-η-κ-μ fading model IEEE Transactions on Antennas and Propagation.

[6] Patzold M 2012 Mobile Radio Channels (Chichester: John Wiley & Sons Ltd) p 11.

[7] 3GPP 2019 Recommendation TR 38.901 V16.1.0 : Study on channel model for frequencies from 0.5 to 100 GHz.

[8] MathWorks 2019 5G New Radio Design with MATLAB (www.mathworks.com/campaigns/offers/5g-technology-ebook.html?elqCampaignId=10588).

[9] Jaeckel S, Raschakowski L and Thiele L 2019 QuaDRiGa - Quasi Deterministic Radio Channel Generator, User Manual and Documentation (Berlin: Fraunhofer Heinrich Hertz Institute).

[10] Quamcom Resarch & Technology AB 2020 (www.qamcom.se/research/5gsim).

[11] Zaidi A, Athley F, Medbo J, Gustavsson U, Durisi G and Chen X 2018 5G Physical Layer: Principles, Models and Technology Components (London: Academic Press, Elsevier).

[12] Zanella A, Verdone R 2012 Pervasive Mobile and Ambient Wireless Communications: COST Action 2100 (London: Springer Science & Business Media).

[13] Nurmela V 2015 METIS Channel Models (https://metis2020.com).

[14] Sun S, McCartney G and Rappaport T 2017 A novel millimeter-wave channel simulator and applications for 5G wireless communications (Paris:IEEE International Conference on Communications (ICC)).

[15] Tan Y 2020 Statistical Millimeter Wave Channel Modelling For 5G and Beyond (Edinburgh:Heriot-Watt University, School of Engineering and Physical Sciences).

[16] Eldowek B, El-atty S, El-Rabaie E and El-Samie F 2017 Second-order statistics channel model for 5G milimeter-wave mobile communications Arabian Journal for Science and Engineering,

[17] Jaksic D, Bojovic R, Spalevic P, Stefanovic D, Trajkovic S 2017 Performance analysis of 5G transmission over fading channels with random IG distributed LOS components International Journal of Antennas and Propagation.

[18] Panajotovic A, Sekulovic N, Bandjur M and Stefanovic M 2017 Second-order measures of performance of dual SC macro-diversity system with unbalanced BSs exposed to CCI in composite fading channels IETE Journal of Research **64**.

[19] Sassan A 2019 5G NR: Architecture, Technology, Implementation and Operation of 3GPP New Radio Standards (London: Elsevier).

IOP Conf. Series: Materials Science and Engineering **900** (2020) 012017 doi:10.1088/1757-899X/900/1/012017

Development of methods for analyzing patterns of current consumption in a system for wireless monitoring the effectiveness of metalworking production

A D Lukyanov*, M A Vernezi, O I Katin, V V Dolgov, I A Zimovnov and S G Studennikova

Don State Technical University, Rostov-on-Don, Russia

*e-mail: alexlukjanov1998@gmail.com

Annotation. Now the world is actively developing a trend of digitization of production enterprises. However, not all existing enterprises are equipped with a proper system for collecting and displaying information. Hence, the task of simple and fast integration of the monitoring system for the technological process produced by Metalworking equipment is quite relevant. To solve this problem, the article presents algorithms for accounting for the number and type of manufactured parts on Metalworking machines, as well as diagnosing the quality of the technological operation, based on temporary patterns of electric current consumed by the machine. This information, collected from the entire production cycle, will allow you to plan production more effectively.

1. Introduction

Monitoring of the production process is one of the essential components of successful and effective production management [11, 12]. Modern CNC machines are usually equipped with built-in monitoring systems that provide comprehensive information about the processing process via a digital channel. However, in the absence of such a built-in service, the problem arises of equipping the equipment with a third-party, independent device that allows you to provide data for the monitoring system of the workshop/enterprise level.

Currently, there are quite a lot of such systems on the market, but almost all of them require some degree of integration with the monitored equipment and / or continuous interaction with the operator.

2. Setting the monitoring task

The aim of monitoring the processing process on CNC machine tools can be divided into three levels (both in terms of the degree of detail of monitoring and the complexity of its implementation).

- the first level of monitoring is the monitoring of the actual operating time of the machine in the part processing mode, in the idle mode, as well as (and) the idle time of the metal cutting machine.

- the second level of monitoring, including the previous one, adds accounting for the type and quantity of parts manufactured on the machine;

- the third level of monitoring, including the previous two, adds diagnostic capability processing, evaluating, changing current magnitude and the duration of individual operations, the correct modes of operation, the consistency of the parameters and properties of specified items, quality and tool wear.

3. Analysis of methods for monitoring the operation of machine tools

The first level of monitoring is implemented by simple means of analyzing the intervals of finding the current value between certain levels, using the simplest methods of filtering and / or statistical data processing, such as digital filtering, time window averaging, median filters, etc. This level is implemented in most existing systems for monitoring the operating time of industrial equipment, for example, counters of operating time. Our method of calculating time allows you to divide the total time of equipment operation by the time of idling and the time of work during the processing of the part. This information allows you to more accurately predict the remaining resource until the next maintenance service, as well as improve the efficiency of production planning.

The implementation of the second level of monitoring requires the use of methods for selecting and recognizing patterns in time series, for which correlation methods [3], synchronous filtering methods [4], methods for analyzing statistical characteristics on the time window, OLS, etc. can be used. A review of the most promising methods that are used to detect and isolate patterns in time series can be found in [13, 14].

The third level of monitoring, in addition to using the methods of the two previous ones, requires taking into account the features of the processing process, namely: physical processes occurring during cutting [18], properties of the processed material, characteristics and wear of the processing tool [15], properties of the coolant, etc. For example, we can use the simple formula of cutting force [19]:

$$F_0 = \rho\left(S_p - X_1\right)$$

where ρ - is the stiffness coefficient of cutting, included all characteristics of material and cutter, and S_p - is a feed per rotation. An increase in the hardness of the material or wear of the cutter will cause an increase in cutting force and, as a result, an increase in the current consumed by the machinetool, which will lead to an increase in the amplitude of the pattern as a whole (in the first case), or only in the processing interval with a worn tool (in the second case) [20].

The full use of this direction of monitoring and diagnostics should be accompanied by the construction of a mathematical model of the processing process [16, 17], taking into account the cutting, friction and dynamics of the machine drives, and its verification using standard workpieces and tools. The recording of the current pattern of the reference processing process in the future should serve as a reference for assessing the status of the processing of subsequent workpieces. In the study, we considered 3 methods of work control of machine tools: identification of the main actuator by means of rotation sensors; identification of the main actuator by measuring the magnetic field around the main drive; measuring the current consumption of the entire machine by means of a current transformer. The first two methods are applicable for tasks that correspond to the first level of monitoring.

Tracking the rotation of the main drive is as follows: on the rotating shaft of the engine, a "mark" is installed, which is either a reflective pad (for optical or infrared sensors), a magnet (for sensors based on a reed switch or a Hall sensor), a metal curtain (for inductive sensors or Hall sensors with a built-in magnet), and so on. A sensor corresponding to the label is installed on the stationary part of the engine, which provides an indication of the fact of rotation of the main drive. The disadvantage of this method is the complexity of mounting this measuring system, and sometimes it is impossible.

Experimental studies on measuring the rotating magnetic field around the main drive stator have shown their good magnetic shielding and the futility of this method of measurement.

The analysis of options for solving the problem showed that the only method that meets the requirements of all three stages of monitoring is to control the electric current consumed by the equipment under study. This method provides the simplest integration into existing systems with no structural changes.

4. Methods of conducting experiments and data analysis

Experiments on measuring the current consumed by industrial equipment were conducted at a Metalworking site equipped with multi-spindle machines and processing centers. The VTL-600 type machine was used as the main one (Fig. 1)

Figure 1. Vertical lathe VTL-600, used for research

An inductive current sensor of the SCT-013-000 type [1] was connected to one of the phases of the input feeder of the machine. The induced current was rectified and averaged using a bridge diode scheme on Schottky diodes, filtered by a second-order passive low-pass filter, and connected to the input of A 10-bit ADC wavgat D1 mini microcontroller [2] based on ESP-8266 (fig. 2). Current values were read every 100 ms, averaged over 1 second, and sent to the cloud server using the MQTT Protocol. Data was stored on the server and then analyzed on the local computer.

a b

Figure 2. Components used in the design of the measuring system.
a - current sensor, b - microcontroller board

IOP Conf. Series: Materials Science and Engineering **900** (2020) 012017 doi:10.1088/1757-899X/900/1/012017

During three months of experiments, more than 850 MB of data was accumulated, in particular, from the machine in question – more than 4 million. current counts, which corresponds to ~1200 working hours. In fig. 3 you can see the developed web-interface to data, collected in cloud. It realize 1st level of monitoring – demonstrate the curve of current consumption, mean current, and intervals between patterns.

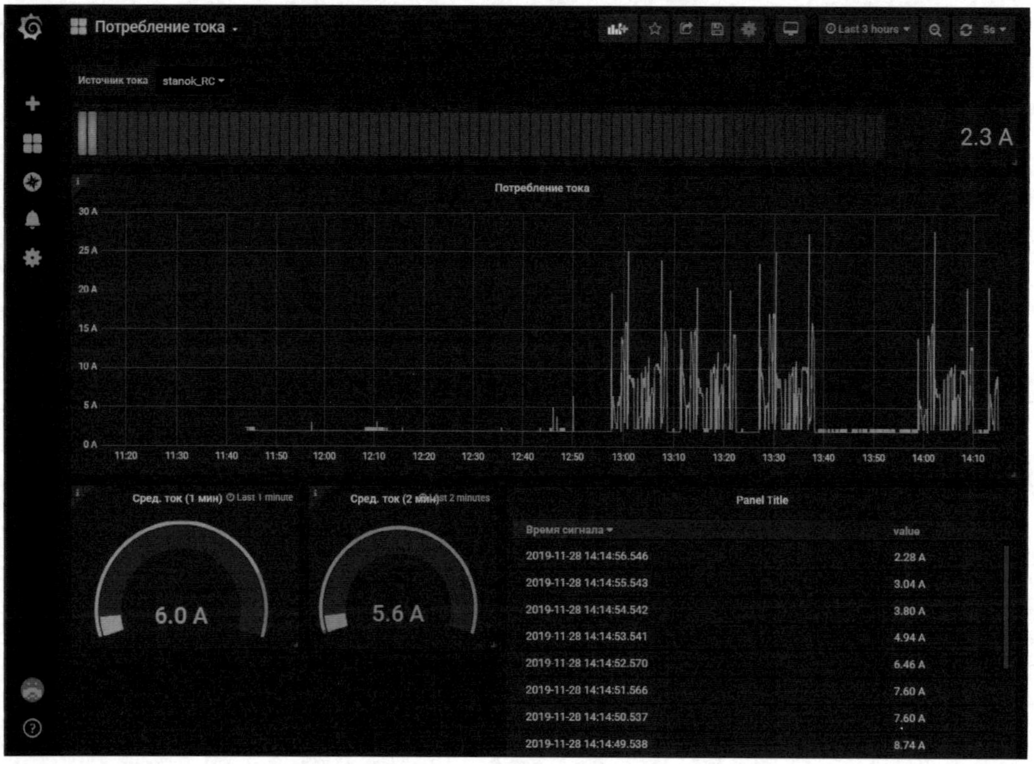

Figure 3. Web – interface to data, collected in clouds

A typical graph of the current consumed by the machine in the process is shown in fig. 4. the graph shows a period of time that is 120,000 seconds away from the beginning of measurements (one frame is 10,000 seconds). In the drawing, you can see the sections of idling (minimum current values in the region of zero and after 5000 seconds), and fluctuations in current consumption that occur during various processing operations.

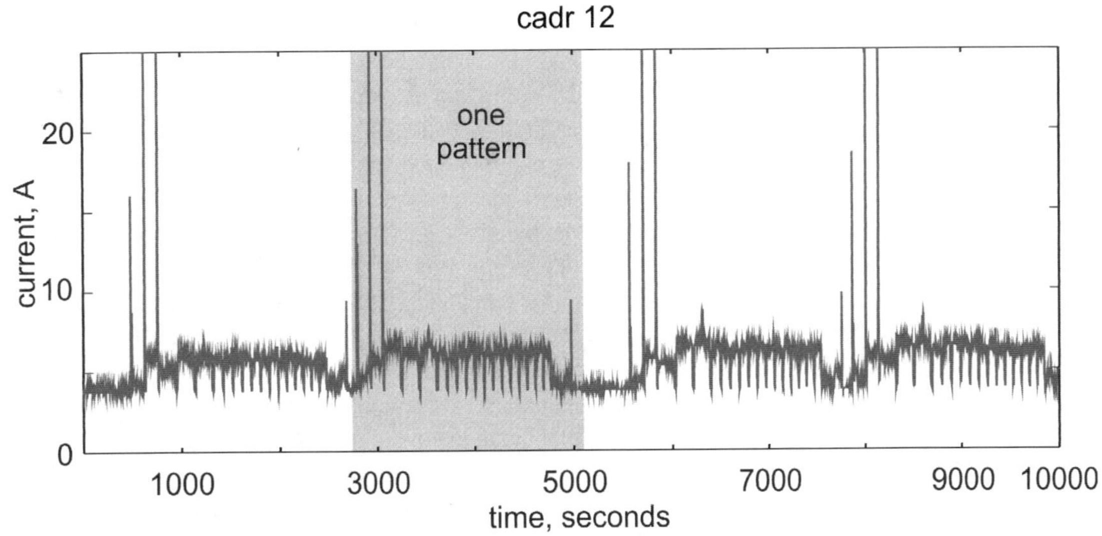

Figure 4. Typical view of the current consumed by the machine during operation, frame No 12

Figure 5 shows, for example, another section of current recording for frame 57. In this case, the part manufacturing operation is shorter, and more of them get into the frame. Areas of downtime are more noticeable.

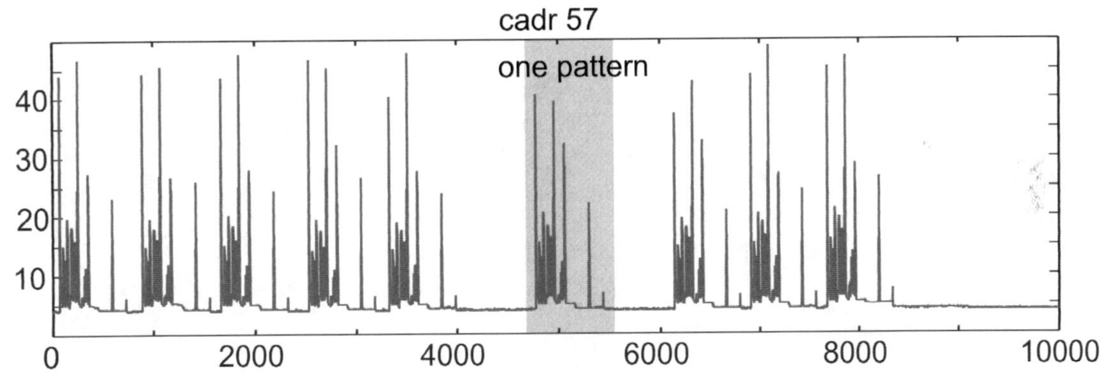

Figure 5. Typical view of the graph of the current consumed by the machine during operation, frame No 57

By analyzing the data obtained, you can divide the task of diagnostics and monitoring of industrial equipment into three hierarchical levels, determined by both the complexity of the analysis [3, 4] and the information received by the end user:

Level 1. Calculating engine hours and analyzing equipment downtime. The problem can be solved almost immediately after the sensors are installed and data accumulation begins by determining the idling level of the machine, and dividing the diagram into sections "work" - "simple". This service is provided by almost all software products on the market, and does not require significant costs when configuring the customer.

Level 2. Counting the parts produced on the machine and sorting them by type. This service requires more in-depth data analysis, associated with the preliminary accumulation of information about the types of manufactured parts and further classification of patterns during the recording of the current value. Correlation analysis methods, synchronous filtering methods, and others are used to find and classify patterns. However, the time distortion of the pattern can be a significant complication when it is stretched, compressed or torn for technological reasons, for example, when a modifier or program stop is introduced into the CNC program. However [5], this task can be successfully solved, but it will require time-consuming initial configuration and further support from the customer. As a rule, common programs for recording the operating time of industrial equipment do not offer such a service.

Level 3 includes a full-fledged diagnostics of the process of Metalworking (or other) production due to an in-depth analysis of the current amplitude in the pattern itself and its comparison with the standard. The idea of the analysis is that the current consumed in the production process depends on both the properties of the processed material and the state of the cutting tool. That is, as a deviation of phys.- mech. both the billet properties from the standard and the tool wear will be displayed in the value of the current consumed [6, 7]. However, such diagnostics are currently not implemented by any of the known commercial software products, because it is labor-and knowledge-intensive, and, in fact, requires disclosure of the cutting dynamics for each individual part being processed.

Analysis of the accumulated experimental material [8] has shown that the above hypothesis is true, and, moreover, tool wear can be detected even for consecutive patterns (fig. 6, 7). Figure 5 shows quite clearly the increase in the amount of current consumed from 1 to 3 pattern, most likely associated with the gradual wear of the tool performing the same type of operation (turning) [9,10]. Figure 6 shows the same increase in the current level from the 1st pattern to the 4th.

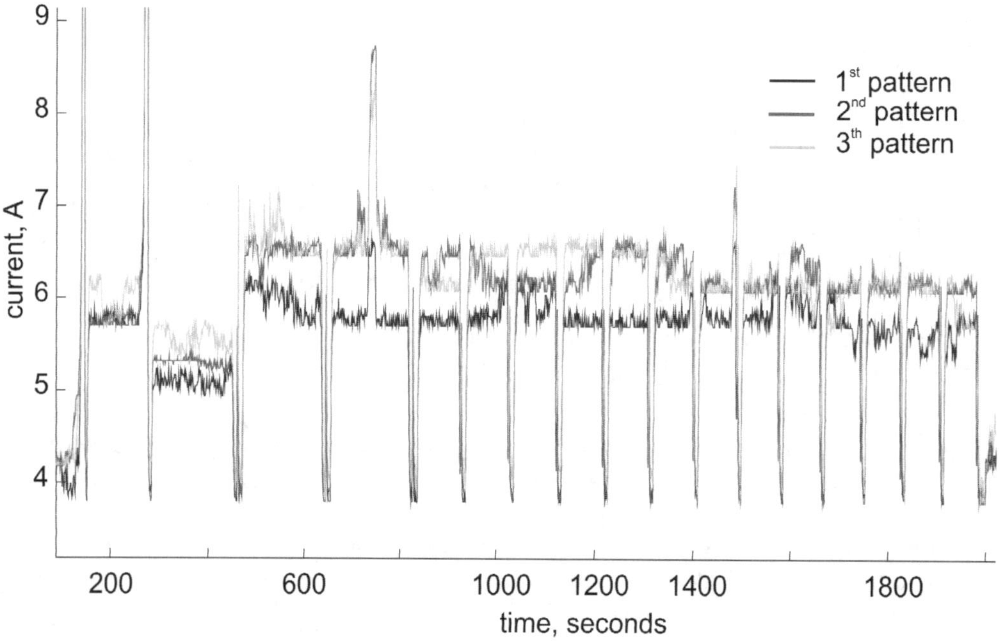

Figure 6. Overlapping sequential patterns, see figure 3

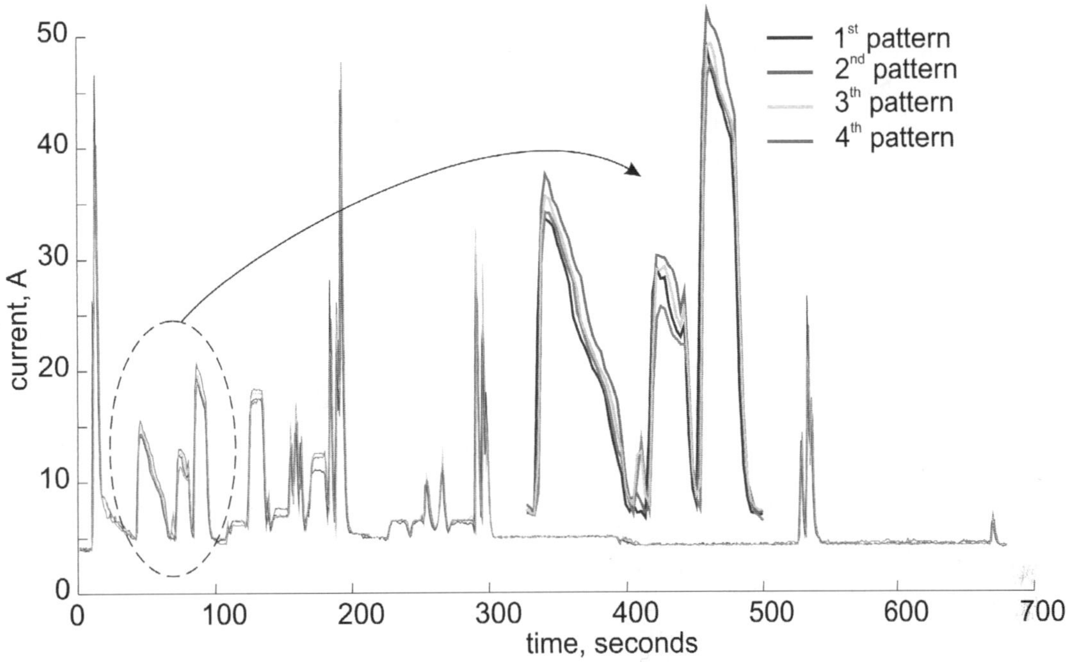

Figure 7. Overlapping sequential patterns, corresponding to Fig. 4.5. callout-an enlarged area in the region of 0-100 seconds

5. Discussion of results

The research presented in this paper is, in fact, an assessment of the feasibility of implementing a system for monitoring power consumption by production equipment, monitoring services, evaluating the effectiveness and diagnostics of the equipment used. The accumulated experimental data allow us to say that with proper analysis (perhaps using machine learning methods and pattern recognition methods), it is possible to create a monitoring and diagnostics system that corresponds to the concepts of "digital production".

Potentially, the current monitoring system of metalworking equipment can serve as the basis for creating a "digital passport" of the manufactured part, since the current pattern corresponding to the manufacturing process reflects the characteristics and parameters of all operations performed on the machine.

Conclusions

According to the results of the research, it can be argued that the approach to creating a system for monitoring and diagnosing the operation of industrial equipment, based on measuring the current consumed by the equipment, and its further analysis and processing, will allow you to implement all three proposed diagnostic levels, bringing this area as close as possible to «digital production».

Gratitudes

The authors express their gratitude to the management and staff of the Shakhty Hydraulic Drive Plant, who provided invaluable support in conducting experimental research.

The studies carried out with support of Fund of assistance to innovation grant "UMNIK".

References

[1] A.Chechin. Current measurement by a sensor SCT-013-030 Arduino.ua. URL: https://arduino.ua/art65-izmerenie-sili-toka-datchikom-sct-013-030 (Access 01.06.2020)

[2] Wemos D1 R2 и mini based on esp8266 Arduino Master, Russian Arduino community. URL: https://arduinomaster.ru/datchiki-arduino/esp8266-wemos-d1-mini-raspinovka (Access 01.06.2020)

[3] A. A. Ovchinnikov, A. E. Khramov, Annika Luttiehann, A. A. Koronovsky, J. van Luzhetaalar. Method of diagnostics of characteristic patterns on observed time series and its experimental implementation in real time in relation to neurophysiological signals. Journal of technical physics, 2011, volume 81, issue 1, pp. 3-10Chiu B., Keogh E., Lonardi S. Probabilistic discovery of time series motifs. *Proceedings of the 9th ACM SIGKDD International Conference on Knowledge Discovery and Data Mining*. August 24-27, 2003. DC, Washington: ACM. 2003. P. 493-498.

[4] Mueen A., Keogh E., Zhu Q., Cash S., Westover B. Exact discovery of time series motifs // Proceedings of the SIAM International Conference on Data Mining (SDM 2009). American Statistical Association (ASA). 2009. P. 473-484.

[5] Mueen A., Keogh E.J. Online discovery and maintenance of time series motifs. *Proceedings of the 16th ASM SIGKDD Conference on Knowledge Discovery and Data Mining (KDD 2010),* July 25-28, 2010. Washington, DC: ACM. 2010. P. 1089-1098.

[6] Patel P., Keogh E.J., Lin J., Lonardi S. Mining Motifs in Massive Time Series Databases *IEEE International Conference on Data Mining.* Dec. 9-12. 2002. DOI: 10.1109/ICDM.2002.1183925.

[7] Ogras Y., Ferhatosmanoglu H. Online summarization of dynamic time series data. The VLDB Journal, January, 2006. P. 84-98.

[8] Palpanas T., Vlachos M., Keogh E. Online Amnesic Approximation of Streaming Time Series // Advances in spatial and temporal databases. 10th international symposium, (SSTD 2007), Boston, MA, USA, July 16–18, 2007. Proceedings P.148-166.

[9] Yankov D., Keogh E., Medina J., Chiu B., Zordan V. Detecting Time Series Motifs Under Uniform Scaling *Proceedings of the 13th ACM SIGKDD International Conference on Knowledge Discovery and Data Mining.* San Jose, California, USA, August 12-15, 2007, DOI: 10.1145/1281192.1281282.

[10] Agrawal R., Lin K., Sawhney H., Shim K. Fast Similarity Search in the Presence of Noise, Scaling, and Translation in Time-Series Databases Proceedings of the 21st VLDB Conference Zurich, Switzerland. 1995. P 200-213.

[11] Advances in Manufacturing, Production Management and Process Control: Proceedings of the AHFE 2019 International Conference on Human Aspects of Advanced Manufacturing, and the AHFE International Conference on Advanced Production Management and Process Control, July 24-28. 2019. Washington D.C., USA.

[12] Advances in Production Management Systems: Innovative and Knowledge-Based Production Management in a Global-Local World: IFIP WG 5.7 International Conference, APMS 2014, Ajaccio, France, September 20-24. 2014. Proceedings, Part II.

[13] Ovchinnikov A.A., Temples A.E., Luttyehann Annika, A Koronovsky.A., J. van Lugetaalar / Method for the diagnosis of characteristic patterns on observables time series and its experimental implementation in the mode real time as applied to neurophysiological signals // ZhTF. 2011. Vol. 81 No. 1 p. 3 (in Russian).

[14] Ponomarev D.I. Using pattern detection algorithms to identify operator gestures in the records of a manipulator control signal. *Transactions of MIPT*. 2012. Vol. 4 No. 2 p. 187-197 (in Russian)

[15] Zakovorotny V.L., Gubanova A.A., Lukyanov A.D. Stability of Shaping Trajectories in Milling: Synergetic Concepts. *Russian Engineering Research*. 2016. Vol. 36. No 11. p. 956-964.

[16] Zakovorotnyi V.L., Lukyanov A.D., Bykador V.S Dynamic Self-Organization in Cutting Process Evolution. 6th International Conference on Mechanics and Materials in Design, M2D 2015 2015. p. 119-134.

[17] Shapovalov VV, Ozyabkin A.L., Kharlamov P.V. Application of Methods of Physical and Mathematical Modeling and Tribospectral Identification for Monitoring Friction Mechanical Systems. *Bulletin of Mechanical Engineering*. 2009. No. 5. P. 49-57.

[18] Zakovorotnyi V.L., Lukyanov A.D. System Synthesis of Machine Tool Manufacturing Process Control Based on Synergetic Conception. *Procedia Engineering* 2016. p. 370-375.

[19] Zakovorotnyi V.L., Lukyanov A.D., Gubanova A.A., Khristoforova V.V. Bifurcation of Stationary Manifolds Formed in the Neighborhood of the Equilibrium in a Dynamic System of Cutting. *Journal of Sound and Vibration*. 2016. Vol. 368. P. 174-190.

[20] Zakovorotnyi V.L., Lapshin V.P., Babenko T.S. Modeling of Tool Wear: Irreversible Energy Transformations. *Russian Engineering Research*. 2018. Vol. 38. № 9. P. 707-708.

IOP Conf. Series: Materials Science and Engineering **900** (2020) 012018 doi:10.1088/1757-899X/900/1/012018

Method for finding the shortest route on an auxiliary weighted digraph when filtering signals

I S Shrayfel, V I Marchuk*, A B Mikhailov, I M Malcev

Don State Technical University, Laboratory of electrochemical and hydrogen energy. Town of Shakhty, Rostov Region, Russia

*e-mail: marchuk@sssu.ru

Annotation. This paper provides an analytical justification and results of practical testing of a new approach to signal processing, which can be considered as a piecewise constant changing function under conditions of a priori uncertainty about the signal parameters. The simulation results presented in this paper confirm the analytical results obtained, and allow us to conclude that the proposed method can be used for processing measurement results under conditions of a priori uncertainty about the useful signal function.

1. Problem statement

Currently, when processing measurement results, it is very important to isolate the signal (useful component) in the presence of an additive noise component in the conditions of a priori uncertainty about the signal function and statistical characteristics of the additive noise component. Classical processing methods are based on the presence of a priori information about the components of the measured process [1]. The less a priori information about the measured process, the more difficult it is to process the measured signals. In this regard, it is very relevant to solve the problem with a minimum amount of a priori information about the parameters of the measured signal [2-9].

The focal idea of This work is to develop a method for reducing the error of selecting a useful signal for a limited amount of a priori information on the example of a piecewise constant function based on the use of the least squares method. Let the resulting signal be an additive mixture of a useful signal and a random noise component, $y(t) = u(t) + \eta(t)$, where $y_1, y_2, \dots, y_n$ values received at time points $t_1 < t_2 < \cdots < t_n$. We assume that the deterministic signal $u(t)$ is a piecewise constant function. In this paper we propose a method for filtering the signal $y(t)$, based on the idea of the least squares method in the absence of a priori information about the parameters of this function.

2. Problem solution

Let $x_1, x_2, \dots, x_n -$ be a sequence of n real numbers. Then there are, and the integer is uniquely defined m, a sequence $(n_k)_{k=0}^{m}$ integers such that

$$n_0 = 0 < n_1 < n_2 < \cdots < n_m = n, \quad (1)$$

and a sequence of pairwise distinct real numbers $z_1, z_2, \dots, z_m$, satisfying the relations $x_i = z_k, i = n_{k-1} + 1, n_{k-1} + 2, \dots, n_k; k = 1, 2, \dots, m$. Counting $x_1, x_2, \dots, x_n$ as the desired approximation of the pure signal, for each number

$1 \le m \le n$ and an arbitrary sequence (1) select the values z_k from the condition for minimizing the amount

$$\sum_{i=1}^{n}(y_i - x_i)^2 = \sum_{k=1}^{m}\sum_{i=n_{k-1}+1}^{n_k}(y_i - z_k)^2. \quad (2)$$

Obviously, this condition reduces by minimizing all internal sums in the right part of the expression (2). In turn, these internal sums are minimal, if and only if $z_k = \bar{y}_k = \frac{1}{n_k - n_{k-1}}\sum_{i=n_{k-1}+1}^{n_k} y_i$, $k = 1, 2, \dots, m$. Therefore, the smallest value of the sum (2) is equal to:

$$\sum_{k=1}^{m}\sum_{i=n_{k-1}+1}^{n_k}(y_i - \bar{y}_k)^2. \quad (3)$$

This shows that minimization of expression (3) over the set of all sequences of the form (1) is meaningless. The result would be the conversion of this amount to zero when $n_k = k, k = 1, 2, \dots, m = n$ as a result, the equalities would be fulfilled $x_i = y_i, i = 1, 2, \dots, n$, without giving any filtering method. Moreover, the ability to perform not all, but at least some of the equalities $n_k = n_{k-1} + 1, k = 1, 2, \dots, m$ casts doubt on the value of minimizing the sum (3) as a means of approximating a pure signal $u(t)$. In this regard, a way out of this provision is to exclude from consideration all sequences of the form (1) that do not meet the requirement

$$n_k \ge n_{k-1} + 2, k = 1, 2, \dots, m. \quad (4)$$

Suppose that the time intervals separating one count from another are small relative to the lengths L_k of the intervals of constancy of the pure signal. Consider the case $t_i = (i-1)h, i = 1, 2, \dots, n$, where $h(> 0)$ – is some given constant. If the smaller of the lengths L_k exceeds only $2h$, at least two points belong to each of the specified intervals t_i, t_{i+1}. Then the restriction (4) is not, in fact, since it is satisfied by sequence (1) the found for $(u_i = u(t_i))_{i=1}^{n}$.

Let's bring the sum (3) to a more convenient form for calculation. Write down

$$\sum_{k=1}^{m}\Big(\sum_{i=n_{k-1}+1}^{n_k} {y_i}^2 - (n_k - n_{k-1})\,{\bar{y}_k}^2\Big) = \sum_{k=1}^{m}\sum_{i=n_{k-1}+1}^{n_k} {y_i}^2 + \sum_{k=1}^{m}(n_{k-1} - n_k){\bar{y}_k}^2 = \sum_{i=1}^{n} {y_i}^2 + \sum_{k=1}^{m}\frac{\left(\sum_{i=n_{k-1}+1}^{n_k} y_i\right)^2}{n_{k-1} - n_k}.$$

Let's give the last amount the form

$$\sum_{k=1}^{m}\frac{(Y_{n_k} - Y_{n_{k-1}})^2}{n_{k-1} - n_k}; \quad (5)$$

where

$$Y_0 = 0, Y_j = \sum_{i=1}^{j} y_i, j = 1, 2, \dots, n. \quad (6)$$

Equality$(3) = \sum_{i=1}^{n} {y_i}^2 + (5)$ allows you to minimize the sum (5) instead of (3), which is reflected in the subsequent algorithmic solution of the problem.

Setting a weighted oriented graph G with these characteristics: 1) the graph contains $(n+1)$ vertices, with assigned numbers 0, 1, 2, … , n; 2) digraph G has an arc (i, j), originating from a vertex with a number i and entering the vertex with the number j, if and only if $i + 2 \le j$; 3) the weight of each arc (i, j) is set by the expression $d_{ij} = \frac{(Y_j - Y_i)^2}{i - j}$. The calculation starts with calculating the values (6) and then the weights d_{ij}. The problem of minimizing the sum (5) is reduced to the problem of the shortest route on the digraph G. Since $d_{ij} \le 0$, it should be noted its applicability to digraphs with weights of any signs. In this case, it is better to talk about the minimum route rather than the shortest one. Supplying G with arcs $(i, i+1), i = 0, 1, \dots, n-1$, endowing them with meaning $+\infty$ in order to prevent these additional arcs from falling into the required minimum route. Let λ_j – length of the minimum route on the digraph G, connecting its vertices 0 и j, $0 \leq j \leq n$. Let's assume that for some $1 \leq j \leq n$ minimum route from 0 up to j already built. Then obviously, $\lambda_j = \min_{0 \leq i \leq j-1} (\lambda_i + d_{ij}), 1 \leq j \leq n$. This recurrence relation allows, starting from the relation $\lambda_0 = 0$, to sequentially calculate the values $\lambda_1, \lambda_2, ..., \lambda_n$. After calculating each value λ_j should fix the number $i = i(j)$, where the specified minimum is reached.

Suppose that, $r_0 = n$ и $r_1 = i(r_0)$, then we'll find the numbers $r_2 = i(r_1), r_3 = i(r_2)$ ... etc. After a finite number of steps, the next value r_m will be zero. Having accepted that $n_k = r_{m-k}, k = 0, 1, ..., m$, get the optimal sequence of numbers $(n_k)_{k=0}^{m}$, satisfying conditions (1) and (4). Now the desired approximation to the signal estimation can be found using the formulas $x_i = \frac{Y_{n_k} - Y_{n_{k-1}}}{n_k - n_{k-1}}, i = n_{k-1} + 1, n_{k-1} + 2, ..., n_k; k = 1, 2, ..., m$.

Below are the results of applying the described filtering method to the same piecewise constant pure signal in the presence of additive normal noise with zero mathematical expectation, which are shown in figure 1.

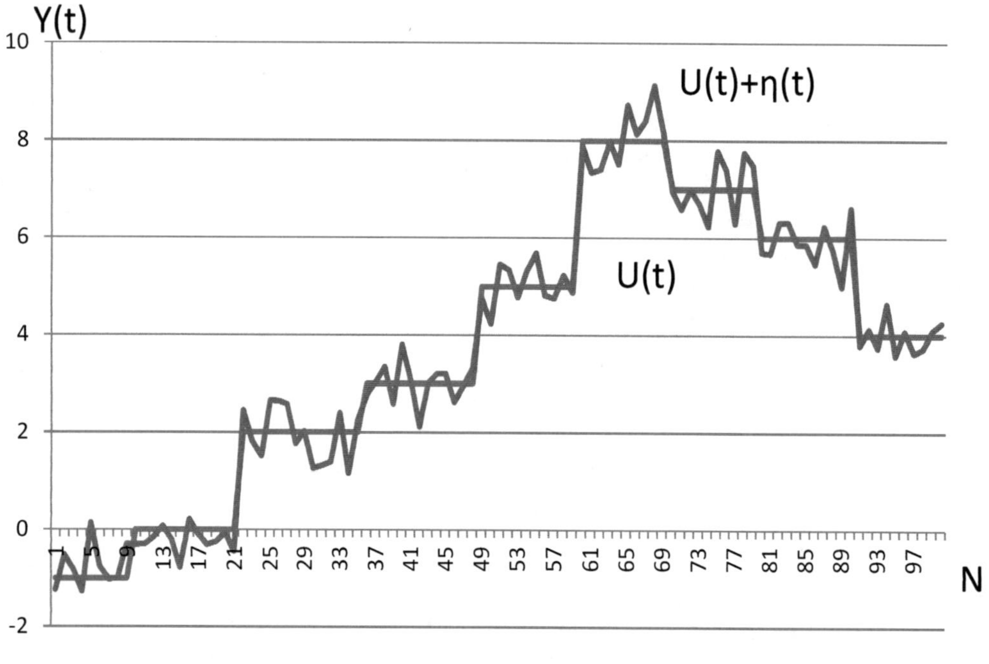

Figure 1. The useful signal model used is U (t) and a noisy signal model U(t)+η(t)

The simulation was carried out with the same model of the useful signal, but the value of the dispersion of the additive noise component changed stepwise and took values of 0.5, 0.75 and 1.0, respectively. the results of processing with the dispersion of the additive noise component $\sigma^2 = 0.5$ are shown in figure 2. When analyzing the results obtained, it should be noted that a priori information about the initial and final coordinates of each stage was completely absent, so is the value of the stage height, i.e. the value of the signal at each of the time points.

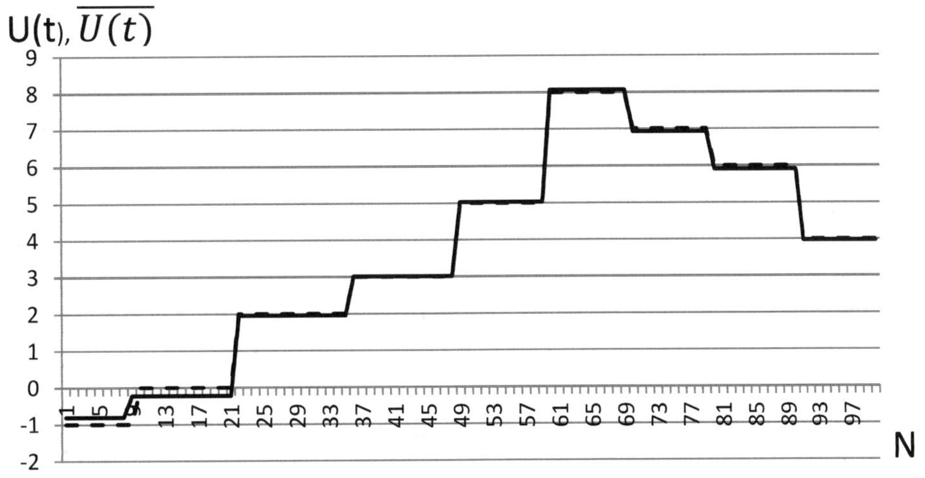

Figure 2. Results of processing the original signal model at $\sigma^2 = 0.5$

The analysis of the simulation results presented in Figure 2 shows that the estimate of the useful signal (dashed line) almost completely coincides with the model of the useful signal (solid line). There is only an error in estimating the amplitude of the useful signal, which is explained by the influence of the structure of the noise component. The initial and final coordinates of the steps of the monotonic function are determined with almost a single probability. However, with increasing noise dispersion, an increase in the approximation error of the original signal model is observed. Figure 3 shows the simulation results for various values of the dispersion of the additive noise component.

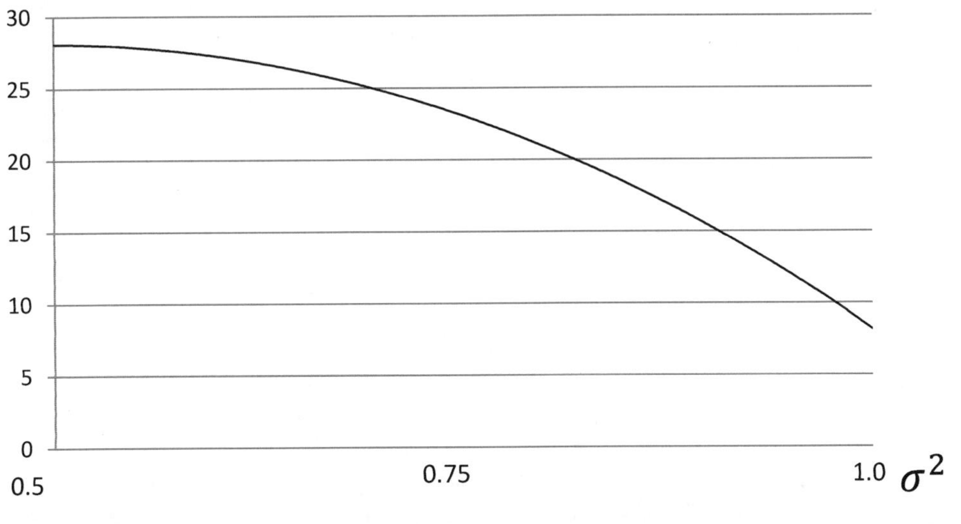

Figure 3. The dependence of the attenuation coefficient of the dispersion of additive noise on the magnitude of its dispersion

An analysis of the results presented in Figure 3 shows that the efficiency of the proposed method decreases with an increase in the dispersion of the additive noise component. So, for example, if $\sigma^2 = 0.5$ it is observed to decrease by almost 28 times, while, when $\sigma^2 = 1.0$ only by 12 times, which is explained by a decrease in the probability of correct detection of the coordinates of each step of the input piecewise constant model of the useful signal function.

Conclusions

1. A new method is proposed for finding the shortest route on an auxiliary weighted digraph when filtering signals.
2. An Optimization of this method using the least squares method has been analytically proven.
3. The simulation results confirmed the theoretical results of the proposed new method of signal filtering in the conditions of a limited amount of a priori information about the signal function.
4. The proposed method allows reducing the dispersion of the additive noise component up to 30 times depending on the values of the input dispersion. However, with increasing dispersion of additive noise, the efficiency decreases.

References

[1] Marchyk V.I., Shrayfel I. S. Methods for isolating the useful component with a priori uncertainty and a limited volume of measurement results. Monograph. – Shakhty: SRSUES, 2008. 163 P.
[2] Marchuk V. I. Improving the reliability of the primary processing of measurement results // Measuring technique. 2003. № 12. C. 3-5.
[3] Marchuk V.I. Increasing the reliability of the primary processing of the results of measurements. Measurement Techniques. 2003. . 46. № 12. 1121-1126.
[4] Marchuk V.., Makov S., Minaev A., Galustov G., Stradanchenko S. Studying of Accuracy of the New Desired Signal Extraction Method by Residuals in a Priory Indeterminacy Conditions. Proceedings of IEEE East-West Design & Test Symposium (EWDTS'2016), Yerevan, October, pp. 343-346, 2016.
[5] Marchuk V.I. Increasing the reliability of the primary processing of the results of measurements. Measurement Techniques. 2003. T. 46. № 12. pp. 1121-1126.
[6] Marchuk V. I., Shrayfel I. S., Minaev A. Y., Buryakov S. A. On optimizing the breakdown of the implementation of a random process in order to evaluate its mathematical expectation. Serbian Journal of Electrical Engineering. 2017. Vol. 14, № 3. pp. 333-342.
[7] Marchuk V. Reducing of noise structure influence on an accuracy of a desired signal extraction. Serbian journal of electrical engineering. 2018. Vol. 15, № 3. pp. 365-370

IOP Conf. Series: Materials Science and Engineering **900** (2020) 012019 doi:10.1088/1757-899X/900/1/012019

Energy-efficient mine ventilation control system

V V Semenov*, A V Fisunov, I O Nikishin

Don State Technical University, Laboratory of electrochemical and hydrogen energy. Town of Shakhty, Rostov Region, Russia

*e-mail: marchuk@sssu.ru

Annotation. The paper considers the principle of constructing a control system of energy-efficient mine ventilation by using new dust detectors that can function in difficult operating conditions. One of such automatic dust meter sensors is described, which contains two sources of radiation in the measuring and reference channels operating at wavelengths in the region of maximum and minimum dust absorption, the radiation which are transmitted sequentially using prisms and mirrors directed through the measuring and reference channels into a single light stream and then to the input of a broadband photodetector. To give the system the possibility to function in difficult operating conditions, the sensor comprises a blower device that contains a protective windows and a heating viewing windows that maintains the temperature of the measuring channel within specified limits. The use of such sensors makes it possible to quickly make decisions about the redistribution of air in mine controlled by the mine ventilation control system.

Currently, in all existing national and international systems for automatic control of mine ventilation, operational air redistribution is carried out according to data on the speed and temperature of air flows on outgoing streams and in working lavas, and dust concentration is usually taken into account indirectly using static characteristics. This fact does not allow you to quickly and correctly make decisions on the redistribution of air in mines. In this regard, the creation of a dust meter for a mine ventilation control system is an urgent task. The purpose of this work is to provide energy-efficient control system of the mine ventilation by using a dust meter with increased accuracy and continuity of operation in the sensor control system.

In most of the existing automated control systems (АТМОС, АКМД, (Russia,) КРАУ (Ukraine), МАЙНОС (England), MCO–060 (Czechoslovakia), BMTA–BMJK (Hungary), A1, K11, K12 (China), CTT 63/40, CTT2, CGA/1 (France) [1, 3, 4], mine ventilation control is carried out not by the dust factor, but by the results of measuring other parameters such as:

- Temperature;
- Speed of the air stream;
- Methane concentration;
- Oxygen concentration.

IOP Conf. Series: Materials Science and Engineering **900** (2020) 012019 doi:10.1088/1757-899X/900/1/012019

Dust control is not carried out in most of the known automatic mine ventilation systems [1, 4]. However, the ventilation control of the mines is necessary to solve the problem of delivering a clean stream of air to the working area and ejecting the dusty stream from it. Therefore, it is advisable to control the ventilation of the mines, taking into account the dust.

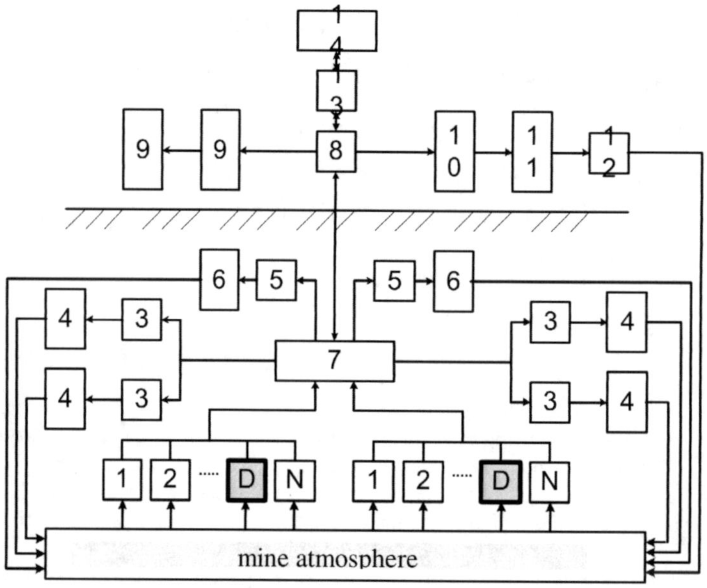

Figure 1. Functional diagram of the mine ventilation control system: 1– methane concentration control equipment; 2– air velocity (flow) control equipment; D – dust meter; N– other sensors; 3 – control equipment of a local air flow regulator with an indicator of the position of the regulatory body; 4 – executive device of the local air flow regulator; 5 – control equipment for a group air flow regulator with an indicator of the position of the regulatory body; 6 – actuator for group air flow regulator; 7 – underground telemechanics; 8 – surface telemechanics; 9 – teleinformation reception devices with recording devices; 10 – air distribution control apparatus; 11 – apparatus for main ventilation fan controller; 12 – actuator for main ventilation fan controller; 13 – matching device; 14 – controlling computer.

Thus, the most important elements of the proposed mine ventilation control system are the sensors that measur the concentration of coal dust. As was shown in a number of works, the operation of such sensors is usually based on optical principles — scattering and attenuation. Dust gauges [5–12] are installed at a distance of 10–15 meters from dust sources (lavas, conveyor lines, etc.) in cleaning drift of coal mine at the incoming and outgoing streams, 10 meters from the pumping and ventilation drifts. The sensors must be fixed at a height corresponding to the miners breathing zone.

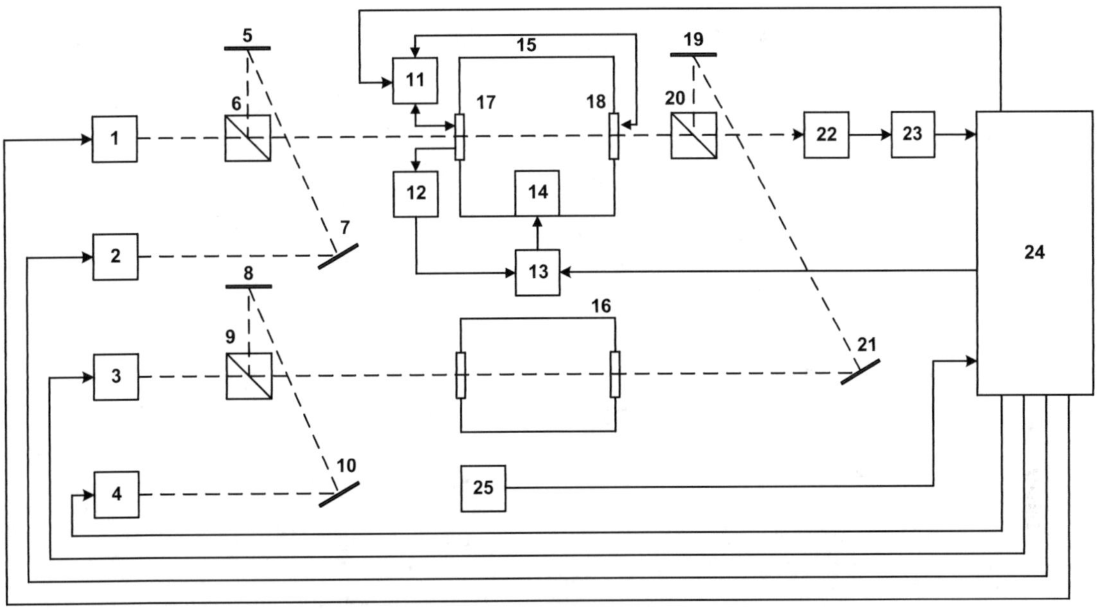

Figure 2. Block diagram of the implemented sensor – dust meter: 1 – 4 – radiation sources, 6, 9, 20 – separation prism, 5, 7, 8, 10, 19, 21 – mirror, 17 – the first protective window, 18 – the second protective window, 15 – measuring channel, 16 – reference channel, 11 – inspection window heating device, 12 – inspection window dust control device, optically connected to the inspection window 17, 13 – airflow control device, 14 – protective airflow device windows, 25 temperature control device, 22 – broadband photodetector, 23 – amplifier, 24– microcontroller.

Figure 2 shows a block diagram of an implemented sensor – dust meter system. Optical sensor - dust meter works as follows [5]. The microcontroller 24 sequentially supplies a pulse voltage to the radiation sources 1.3 and 2.4, while the radiation sources 1.3 have a wavelength in the region of maximum dust absorption, and the radiation sources 2.4 have a wavelength in the region of minimum dust absorption. In this case, the microcontroller sequentially supplies identical bursts of pulses to the emitters 1, 3, 2 and 4. The radiation source 1 sends the generated pack of pulses through the separation prism 6, the radiation enters the measurement channel 15, by the time that the second pack of pulses supplied to the radiation source 3 the radiation arrives at the reference channel 16, then by the time that the third pack of pulses supplied to the radiation source 2 the radiation is transmitted through mirrors 7, 5 and the separation prism 6 is received in the metering channel 15, then by the time that the fourth pack of pulses supplied to the radiation source 4 the radiation is transmitted through the mirror 10, 8 and the separation prism 9 is supplied to the reference channel 16. Next, the process of formation of bursts of pulses to radiation sources of 1.3, 2.4 is repeated. Thus, at the output of the second viewing window 18 of the measuring channel 15, there is a sequence of bursts of light pulses generated by sources 1,2, and at the output of the second viewing window of the reference channel 16 there is a sequence of bursts of light pulses formed by sources 3,4. Next, the light pulses from the radiation sources 1,2 of the measuring channel 15 pass through the separation prism 20 to the input of the photodetector 22. In this case, the pulsed light radiation passing through the measuring channel 15 is attenuated by dust according to the Beer–Lambert law.

Light pulses from the radiation sources of the 3,4 reference channel are transmitted through mirrors 21, 19 and the dividing prism 20 as well as to the input of the photodetector 22. In this case, the light pulses are arranged in the sequence 1, 3, 2, 4. the sequence of electric current pulses Formed by the photodetector is transmitted through the amplifier 23 to the microcontroller 24, where the received sequences are processed. First, the pack formed by the radiation source 1 is subtracted from the pack formed by the radiation source 3 and the pack formed by the radiation source 2 is subtracted from the pack formed by the radiation source 4, then the logarithm of the ratio of the obtained values is determined, which characterizes the level of the measured dust. All operations are performed synchronously.

The device for heating the viewing Windows 11 maintains the temperature of the measuring channel viewing Windows in the range of 210-250OC. The mode of operation of this device is set by the microcontroller 24.

The dust control device for the inspection windows 12 controls the blower 14 with specially cleaned brushes fixed on the fan blades, automatically approaching the inspection windows when the fan is operating. When a certain dust concentration threshold is reached, a light beam reflected at an angle of 135 degrees to the axis of radiation enters the device 12, which is a photodiode, the voltage from which is supplied to the fan control device 13, the operation mode of which is set by the microcontroller 24.

The temperature control device 25, made in the form of a semiconductor temperature sensor and an amplifier, continuously measures the air temperature of the working area and is connected to the microcontroller 24 to correct the fluctuations in ambient temperature.

Thus, the proposed optical sensor - dust meter operates in automatic mode and has design features that allow it to be used for energy-efficient control of the ventilation as part of automatic ventilation control systems. Due to the use of two wavelengths in the region of maximum and minimum dust absorption, the sensor dust meter can improve the accuracy of determining the total dust concentration and eliminate errors that occur when changing uncontrolled parameters: humidity, concentration of carbon dioxide, methane, etc., which makes it possible to quickly and correctly make decisions about air redistribution in mines by the mine ventilation control system.

References

[1] Karpov E.F. and Basovskiy B. I. Coal mines ventilation and degassing control: Reference Guide. M.: Nedra, 1994. 336 P.:

[2] Hygienic requirements for coal enterprises and organization of work. Sanitary rules and regulations (СанПиН 2.2.3.570–96). M. 1996. 84 P.

[3] Vasilenko V. I. Automated ventilation system for non-gas coal mines. Novocherkassk.: NGTU, 1996 172 P.

[4] Pozdnyakov G. A. Scientific basis, methods and technical means of normalizing the atmosphere of the preparatory drifts of coal mines by dust factor: Abstract of PhD thesis. Moscow, 1997. 36 P.

[5] Semenov V. V., Danilenko I. N. Optical dust meter RF Patent № 2558278, IPC G01N21/59, G01N12/15, applicant and patent holder: «Don State Technical University» (DSTU) № 2014106501/28; published 27.07.2015

[6] Semenov V. V., Boyko A. P. and Rumyancev K. E. Optical dust meter for enterprise ventilation control system RF Patent № 2210070, IPC G01N21/59, 21/15 applicant and patent holder South Russian State University of economics and service. № 2001106780/28; published 2003. 3 P.

[7] Semenov V. V. and Popov E. K. Optical dust meter RF Patent № 2510497, IPC G01N21/94 applicant and patent holder South Russian State University of economics and service № 2012132349/28; published 27.03.2014.

[8] Semenov V. V. and Popov E. K. The method for determining the concentration and average particle size of dust RF Patent № 2510498, IPC G01N21/94, G01N 15/02 applicant and patent holder

South Russian State University of economics and service №2012132356/28; published 27.03.2014.

[9] Semenov V. V., Asstaturov Yu G, Khanzhonkov Yu B and Kochkovaya N. V. Photovoltaic method for determining the average concentration and average particle size of dust RF Patent № 2686401, IPC G01N21/94, G01N 15/0205 applicant and patent holder: «Don State Technical University» (DSTU) – № 2018133322; published 25.04.2019.

[10] Semenov V. V., Asstaturov Yu G, Khanzhonkov Yu and Kochkovaya N. V. Optical dust meter RF Patent № 2691978, IPC G01N21/94, G01N 15/0205 applicant and patent holder: «Don State Technical University» (DSTU) №2018133318; published 19.06.2019.

[11] Semenov V. V., Asstaturov Yu G and Khanzhonkov Yu B Optico–Electronic dust meter. *Measurement Techniques*, 2019. № **1**. pp 40-48.

[12] Semenov V. V., Asstaturov Yu G and Khanzhonkov Yu B Optico–Electronic dust meter. *Measurement Techniques* June 2019, Volume **62**, Issue 3, pp. 249–253 | Cite as (https://doi.org/10.1007/s11018–019–01612–2)

IOP Conf. Series: Materials Science and Engineering **900** (2020) 012020 doi:10.1088/1757-899X/900/1/012020

Design and technological support for manufacturing a PAA emitter based on a 3D-model using SolidWorks CAD

V A Lebedev[1], A S Kukareko[1,*] and V A Kukareko[2,3]

[1] Department of mechanical engineering technology, Don State Technical University, Rostov-on-Don, 344000, Russia

[2] Department of industrial safety, Don State Technical University, Rostov-on-Don, 344000, Russia

[3] Junior researcher, Rostov branch of the Russian customs Academy, Rostov-on-Don, 344002, Russia

*e-mail: kukareko.lesha@yandex.ru

Abstract. This article presents a methodology for manufacturing the S-band phased antenna array (PAA) emitter based on a calculated 3D-model, which provides the necessary design and technological requirements. As a method of manufacturing the emitter, the technology of milling processing on a CNC machine was chosen. Verification of milling processes based on the SolidWorks CAM software package was performed.

1. Introduction

In most cases, phased antenna array S-band frequencies are a set of several symmetrically arranged basic elements of the grid, consisting of receiving and transmitting modules (TM), digital signal processing devices, power supply, etc. The fundamental elements in this system are TM, which transmit a signal to the surrounding space, and receive a useful signal from the detection object in a given sector, limited by the directional diagram of the PAA.

TM is made on the basis of a printed circuit Board with antenna emitters fixed on it, using plug-in connectors or soldering. It is the antenna emitters that are of the greatest interest from the point of view of their manufacture, since they have special requirements for shape, processing accuracy and manufacturability [1].

At the moment, in radar engineering, there are quite a large number of different types of antenna emitters that differ in their functional purpose, respectively, design, shape and size. For S-band PAA, symmetrical electric emitters made of aluminum alloys with straight or inclined arms are most often used, depending on the required characteristics. A simplified view of symmetrical emitters with straight and slanted shoulders is shown in figure 1.

Figure 1. Symmetrical PAA emitters.

2. Calculated 3D-model of the emitter

Let's consider an example of a calculated 3D-model of a symmetrical antenna emitter with straight shoulders, which has linear polarization and a wide 90° directivity diagram in the azimuthal plane relative to the main axis of the emitter (Figure 2). It should be noted that the presented 3D-model is ideal, because it fully provides all the necessary tactical and technical characteristics of a given PAA [2]. Overall dimensions of the design emitter — shoulder span 94.6 mm, base width 21 mm, shoulder width 14.5 mm, height 45.5 mm, structure height 41.5 mm, gap between structures 4 mm.

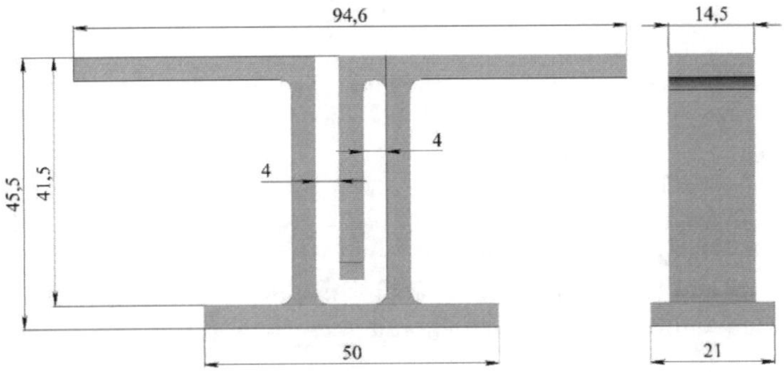

Figure 2. The calculated 3D-model of a symmetrical emitter.

3. Design development of a 3D model of the emitter

At the stage of converting the calculated 3D-model of the emitter into a design model, it is necessary to take into account the possible options for manufacturing the emitter (processing technology), ensuring the strength of the structure, the nature of the electrical installation, and determining the attachment points.

3.1 Selection the emitter material

Aluminum deformable alloy AMg6 in accordance with SIST 4784-2019 was selected as the material for manufacturing the emitter. Aluminum alloy AMg6 has a number of advantages that make the design of the PAA emitter the most reliable, light and technological, among them: low specific gravity; corrosion resistance; vibration resistance; plasticity; low electrical resistivity [3].

3.2 Selection of manufacturing technology

When selecting the manufacturing technology of a complex part, it is necessary to take into account its shape, size and the required accuracy of surface treatment. For the emitter shown in figure 2, it is rational to use one of the two processing technologies:

1. Milling. Processing of flat surfaces and the workpiece contour on the milling machine with CNC;
2. The combined technology of milling and electrical discharge machining. Processing of flat surfaces on a CNC milling machine, contour processing on an electric erosion machine.

Often, the design of PAA uses a fairly large number of emitters, which significantly increases the cost of the product and increases its production time when using combined technology. Thus, with large production volumes, it is rational to use the technology of milling flat surfaces and the emitter contour.

Milling technology involves the use of appropriate tools — milling cutters with a cylindrical structure. Therefore, in the corners of the part, it is necessary to provide rounding of the minimum possible radius, since their excessive increase leads to a deterioration in the coordination of the wave resistance of the antenna and the feed feeder line [1].

Based on the above, the processing of the emitter contour must be performed in two stages, which allows you to reduce the milling depth to half the depth of the workpiece [4]. The minimum radius of rounding in the corners of the part is defined by the following expression (1):

$$R = H/k, \tag{1}$$

H – the milling depth, mm;

k – coefficient, equal to 6, for aluminum, copper alloys and plastics.

Thus, for the emitter shown in figure 2, taking into account the height of the workpiece 23 mm and finishing contour processing of the conditional joint to a depth of 1 mm, the milling depth is equal to 12 mm. From expression (1), the minimum radius of rounding at the corners of the emitter is 2 mm. Therefore, for finishing contour processing, it is necessary to choose a tool with a diameter of no more than 4 mm and a height of the working part of no less than 12 mm.

3.3 Electrical installation and mounting points

Installation of the feeding feeder line is carried out through a connector type SRG50-751FV, for this purpose, the design of the emitter must provide a threaded hole M6×0.75, located on the Central axis of the emitter. The connector is installed in accordance with VR0.364.049 TC, while the supply core is soldered to the central structure of the emitter [5].

To install the emitter on the PAA, there are 4 through-threaded holes M4, which ensure that the emitter is fixed in a given position on the underlying surface of the PAA.

The design of a symmetrical PAA emitter with straight shoulders, taking into account the manufacturability of manufacturing, the requirements for installation in the overall layout of the PAA, is shown in figure 3.

Figure 3. Design 3D-model of a symmetrical PAA emitter

4. The strategy of processing the emitter

To determine the plan for processing parts by milling, software complexes with a CAM-system are used, which allow virtually generating technological transitions, analyzing tool paths, evaluating deviations of the processed surfaces, etc. The algorithm for working in the CAM system includes three stages [6]:

Stage 1. Geometry selection:

- selection of geometric elements of the model to be processed (working, processed, and controlled elements);

- determining the size and shape of the workpiece based on the actual allowance;
- determining the position of the workpiece relative to the zero point;

Stage 2. The selection of a strategy and tool, and the setting of processing parameters:

- selecting a processing strategy (rough, finish, 2D-geometry; 3D-volume geometry; radial geometry; contour geometry, etc.)
- purpose of cutting tools;
- the definition of the cutting modes;
- generation of processing paths.

Stage 3. Backplot and verification:

- checking the correctness of created paths (Backplot function).
- solid-state verification of the workpiece material removal process.

To determine the processing plan for a symmetrical PAA emitter, use the SolidWorks CAM software package. We will perform a technological analysis of this detail based on the above algorithm for working in the CAM-system.

4.1 Plan for milling the emitter

Let's define the geometric elements of the emitter to be processed in accordance with the first stage of work in the CAM-system. The sequence of surface treatment of the part is shown in figure 4.

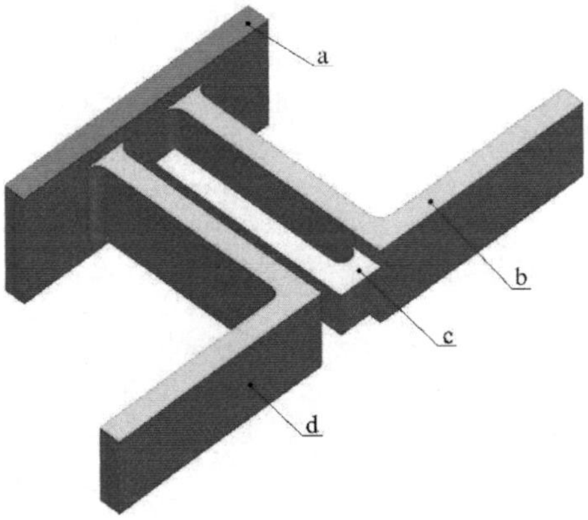

a) the base flat surface;
b) flat surface, milling depth 3.25 mm;
c) flat surface, the milling depth 6.5 mm;
d) contour surface treatment, milling depth 12 mm.

Figure 4. Sequence of processing geometric data emitter elements

To determine the size of the workpiece, it is necessary to rely on two main parameters, namely the overall dimensions of the emitter (94.6×45.5×21) and the actual allowance (min 1 mm, to ensure the 5th class of surface cleanliness) [4]. Based on the design of the emitter, the deviation tolerance of its surfaces relative to the calculated 3D-model, and the actual allowance, the overall dimensions of the workpiece are 97×48×23 mm.

4.2 Strategy for surface milling of the emitter

Determine the processing sequence of the emitter and assign the cutting tool. The CAM-system provides a function for selecting a machine and adding a tool basket. The type of milling machine is selected based on the complexity of the part shape and the number of surfaces to be processed.

To manufacture the emitter will take the standard type of machine provided CAM system, namely a 2.5-axis machine Mill – Metric, class capacity Medium duty, with a maximum flow 16500 mm/min, spindle speed of 12000 rpm [6].

The tool crib is a set of different types of milling cutters that are set by the operator depending on the required part processing strategy. The CAM-system also provides a wide selection of different milling tools that can be changed after generating transitions, depending on the required accuracy and shape of the part.

Now let's use the feature for automatic recognition of geometric elements in a solid-state model, called element extraction. Depending on the previously planned part processing sequence, we make the necessary corrections to the element extraction program using the element recognition functions and processing the part perimeter to the specified milling depth. In this way, the CAM-system determines the emitter processing plan (Figure 5).

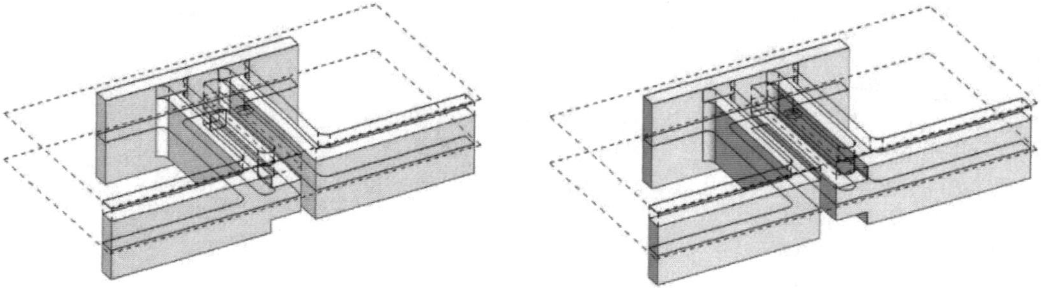

The first milling operation The second milling operation

Figure 5. The treatment plan emitter

To create an emitter surface treatment plan, the CAM-system uses the processing plan generation function, which creates transitions for all processed surfaces based on the rules set in the Technology Database [6]. Note that the presented transitions are preliminary and require correction in terms of their logical combination, replacement of the cutting tool and/or its feed rate, allowance for roughing, changing the sequence of transitions and/or removing them. Thus, to ensure the manufacturability of the emitter, we will correct the transitions generated by the CAM-system.

First, we will replace the proposed cutting tool depending on the processing strategy of the emitter, as well as the cutting speed *V*.

Secondly, we will conduct a technological analysis of preliminary transitions to combine them and correct them. To avoid tool breakage during processing, we will reduce the tool feed rate by 30% during angular movement [4]. Thus, the plan for surface treatment of a symmetrical PAA emitter acquires the structure shown in table 1.

Table 1. Plan of surface treatment of the emitter and the purpose of the cutting tool.

№	Milling strategy	Type of milling tool	Cutting speed *V*, mm/min
	The first milling operation		
1	Rough flat-contour processing of base planes	End mill, 6×24×68 4FL	220÷250
2	Finishing flat-contour processing of base planes	End mill, 4×14×51 2FL	130÷150
3	Rough flat-contour processing of the groove 1	End mill, 6×24×68 4FL	180÷200
4	Finishing flat-contour processing of the groove 1	End mill, 4×14×51 2FL	130÷150
5	Rough flat-contour processing of the groove 2	End mill, 6×24×68 4FL	180÷200
6	Finishing flat-contour processing of the groove 2	End mill, 4×14×51 2FL	130÷150
	The second milling operation		
1	Rough flat-contour processing of base planes	End mill, 6×24×68 4FL	220÷250
2	Finishing flat-contour processing of base planes	End mill, 4×14×51 2FL	130÷150
3	Rough flat-contour processing of the groove 3	End mill, 6×24×68 4FL	180÷200
4	Finishing flat-contour processing of the groove 3	End mill, 4×14×51 2FL	130÷150

4.3 The tool path generation function.

To analyze tool paths for each planned transition, as well as to change the tool entry points, pass coordinates, or cutting tool, use the tool path generation function. Some tool paths in accordance with the processing plan are shown in figure 6.

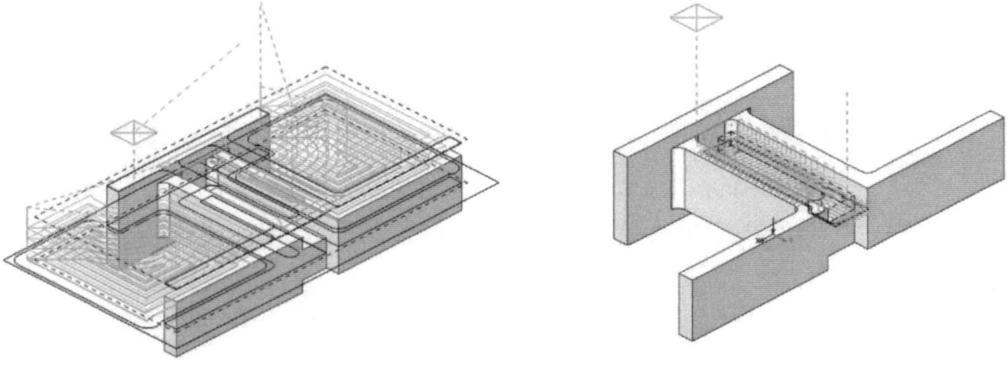

Figure 6. The trajectory of the cutting tool.

4.4 The backplot and processing verification

To confirm the correctness of the created processing plan, the backplot and verification functions are performed. The backplot function is designed to detect errors when processing part surfaces, such as a tool colliding with a part/work piece, a tool holder colliding with a part/work piece, or a mandrel colliding with a part/work piece. Verification is intended for simulation and graphical verification of the material removal process based on the shape of the workpiece and cutting tool.in the CAM-system, it corresponds to the trajectory simulation function. Figure 7 shows some simulations of the cutting tool trajectory when processing the emitter surfaces.

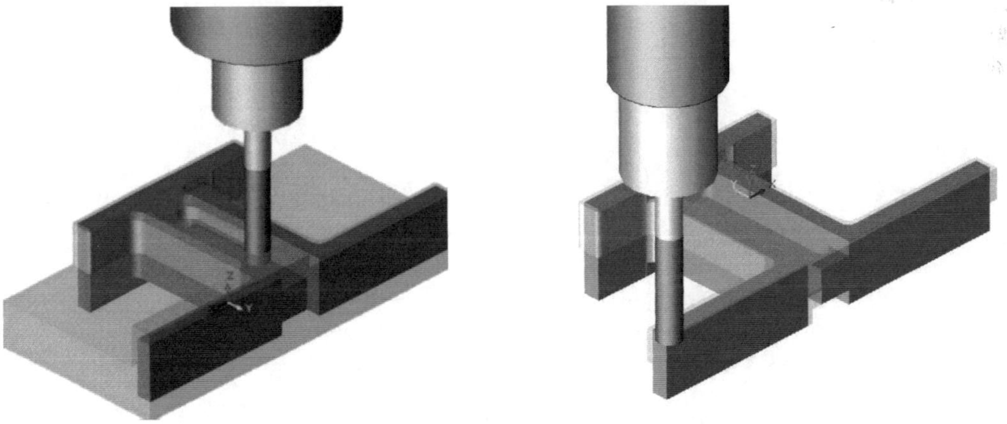

Figure 7. Simulation of the cutting tool trajectory using the verification function

After verification, the programmer-technologist automatically creates a control program (CP) based on G- M-codes using the CAM-system. Then the CP is loaded on a CNC milling machine [6].

5. Conclusion

Thus, this article presents a methodology for manufacturing a symmetrical S-band PAA emitter based on a 3D-model that provides the necessary design and technological requirements. The method of manufacturing the emitter using the technology of milling on a CNC machine is determined, which allows reducing the cost and production time of the emitter in comparison with the often used combined technology of milling and electroerosion processing. To confirm the applicability of the milling technology, verification of the surface treatment processes of the emitter was performed on the basis of the SolidWorks CAM software package.

References

[1] Kubanova V P 2016 *Fundamentals of the theory of antennas and radio wave propagation: study guide* (Samara: Volga state University of telecommunications and Informatics) p 258
[2] Krotova E I 2013 *Fundamentals of design and production technology of RES: study guide* (Yaroslavl: Yaroslavl state University named after Demidov) p 192
[3] SIST 4784-2019. Aluminium and wrought aluminium alloys. Grades: approved and put into effect by the Federal Agency for technical regulation and Metrology on July 31 2019 paragraph N 435
[4] Vereina L I 2017 *Milling technology: study guide* (Rostov-on-Don: Phoenix) p 187
[5] MD 107.434511.001-90. Radio frequency connectors. Selection guide: put into effect by the Directive letter of the SSTU of 23.11.90 N 017-107/K/2622.
[6] Lovigin A A and Teverovskiy L V 2015 *Modern CNC machine and CAD / CAM-system* (Moscow: DMK Press) p 280

AUTHOR INDEX